Advancing Global Bioethics

Series Editors
Bert Gordijn, Dublin City University, Dublin, Ireland
Henk A.M.J. ten Have, St. Vincent's Health Centre, Duquesne University, Pittsburgh, USA

For further volumes:
http://www.springer.com/series/10420

United Nations
Educational, Scientific and
Cultural Organization

UNESCO Chair in
Bioethics and Human Rights
Roma, Italy

The UNESCO Chair in Bioethics and Human Rights (www.unescobiochair.org) organized a workshop on "Bioethics, Multiculturalism and Religion" and the papers submitted here form the collection of this edited volume.

This UNESCO Chair was established in 2009 with two Roman universities. It offers a university framework of reflection and study, providing information and fostering the application of bioethical principles in science, medicine and new technologies based on the Universal Declaration on Bioethics and Human Rights. Through integral education, research and information, it seeks to contribute toward the recognition and promotion of a global and integral vision of bioethics, bringing to light universal values and principles as well as social and legal implications in relation to human rights. By creating a community of persons who are interested in these values, it seeks to promote cultural dialogue and encounters in a spirit of solidarity.

Joseph Tham • Alberto Garcia • Gonzalo Miranda
Editors

Religious Perspectives on Human Vulnerability in Bioethics

Springer

Editors
Joseph Tham
School of Bioethics
Ateneo Pontificio Regina Apostolorum
Rome
Italy

Gonzalo Miranda
School of Bioethics
Ateneo Pontificio Regina Apostolorum
Rome
Italy

Alberto Garcia
UNESCO Chair in Bioethics
and Human Rights
Rome
Italy

Permission has been granted for the reprinting of several articles published previously in *Studia Bioethica* based on original unrevised papers of the workshop. The five papers are as follows: Gonzalo Miranda, Come nacque il "principio di vulnerabilità," *Studia Bioethica* Vol 4, No 2 (2011), 5–7; Martha Tarasco, Vulnerabilidad: Consideraciones sobre la utilización pertinente del término en Bioética, *Studia Bioethica* Vol 4, No 2 (2011), 8–14; Stamatios Tzitzis, Ethique et aspects légaux de la vulnérabilité dans la perspective chrétienne, *Studia Bioethica* Vol 4, No 2 (2011), 15–20; Prakash Desai, Between Tradition and Modernity: Bioethics, Vulnerabilities and Social Change, *Studia Bioethica* Vol 4, No 2 (2011), 21–32; Jonathan Chan, Health care and human vulnerability: A Confucian perspective, *Studia Bioethica* Vol 4, No 2 (2011), 33–40. The original papers of Miranda, Tarasco and Tzitzis were published in their original languages of Italian, Spanish and French in *Studia Bioethica*. All the papers published here in this volume have undergone revisions compared to the earlier versions published in the journal.

Copyediting by Peter Mango.

ISSN 2212-652X ISSN 2212-6538 (electronic)
ISBN 978-94-017-8735-2 ISBN 978-94-017-8736-9 (eBook)
DOI 10.1007/978-94-017-8736-9
Springer Dordrecht Heidelberg New York London

Library of Congress Control Number: 2014940970

Printed on acid-free paper

Springer is part of Springer Science+Business Media (www.springer.com)

In memory of Marion Milan-Lambert who, having felt vulnerable herself, acted on behalf of the most vulnerable, and whose generous gift she bequeathed helped to make this workshop and book project a reality.

Preface

Director, UNESCO Chair in Bioethics and Human Rights

A silent and invisible power is present when people from different backgrounds and belief systems gather in an environment of mutual respect while searching for what they have in common.

Differences among peoples—and of course among scholars—are a matter of fact, but none of them should weaken the universal conviction of the need for and the value of "unrestricted pursuit of objective truth and the free exchange of ideas and knowledge." In modern and contemporary society this assertion could be easily labelled as intellectual fundamentalism. Few would suspect that such an admission and statement would emanate from the Preamble of the Constitution of UNESCO, adopted in London on November 16, 1945.[1] As an international law in vigour, this instrument should not be ignored or misinterpreted. Rather, it is crucial to read it often, to keep it in mind, and to consider it while constructing public policies and making ethical decisions. It should also serve as an important point of reference in our university endeavours.

Hence, neither contemporary relativism nor individualistic subjectivism satisfactorily accomplishes the duty of furthering universal respect for justice as indicated in Article 1 of the UNESCO Constitutional text. In the field of bioethics, "the importance of cultural diversity and pluralism should be given due regard. However, such considerations are not to be invoked to infringe upon human dignity, human rights and fundamental freedoms." (UNESCO Universal Declaration on Bioethics and Human Rights, adopted by acclamation on October 19, 2005)

In other words, cultural diversity and pluralism are not ultimate considerations when discerning about ethics and justice. Respect for diversity and pluralism does not take precedence over the principles that must serve as the building blocks, namely the recognition, protection, and guarantee of everyone's human rights.

The more knowledge we assimilate, the more difficult it is to communicate, share and persuade others of our ideas, convictions and beliefs, especially when confrontation is judged as the only way to reach unity and peace. Creating a simu-

[1] http://portal.unesco.org/en/ev.php-URL_ID=15244&URL_DO=DO_TOPIC&URL_SECTION=201.html.

lation of uniformity can wrongly be considered necessary to govern peoples; in a misguided concern for harmony in a globalized and multicultural world, disdain for the act of dissent can emerge.

To learn from peoples representing other cultures, intellectual integrity and a firm desire of being open are required. I'm convinced that meeting scholars and diverse thought leaders illuminate our perception and knowledge of humankind.

It's easier to remain in the comfort zone of our field of knowledge and not be challenged in our intellectual, moral and religious convictions. But this is neither intellectually truthful nor ethically sound. The UNESCO Constitution declares, "that the wide diffusion of culture, and the education of humanity for justice and liberty and peace are indispensable to the dignity of man and constitute a sacred duty which all the nations must fulfil in a spirit of mutual assistance and concern." This is not only a political responsibility to be considered by nations and States, but also the personal duty we all should undertake as a moral commitment.

Differences could be considered as a deterrent to working together in a spirit of respect. Diversity creates fear everywhere. Our ignorance about what different countries and cultures think and believe often build up walls of suspicion and mistrust. Different beliefs, different ideas, different philosophies, different languages, different colours: too many differences for some people's thoughts and fears to handle.

But a different and more promising way of thinking is possible. Building intellectual and moral bridges is a moral imperative nowadays. It demands a personal and communitarian effort to overcome ignorance of neighbours near and far, since "… ignorance of each other's ways and lives has been a common cause, throughout the history of mankind, of that suspicion and mistrust between the peoples of the world through which their differences have all too often broken into war." (Constitution of UNESCO, Preamble)

A spirit of mutual understanding and friendly dialogue is not only possible, it is also a moral demand that would help to build unity where unity is most needed for the sake of justice and ethics.

Some people could think that religious convictions are an obstacle to scientific and technological progress, and therefore believers—though intellectually competent—should be considered under a lesser light since they could stop or delay new discoveries or therapies due to moral reflections and convictions they bring into consideration. From this unfair assumption emerges fear and a subtle discrimination that should be avoided. Our diverse and globalized world needs to assume the risk of engaging with each other through sincere intellectual efforts. Virtue can spur us on to overcome what divides us to our detriment.

In the field of bioethics, the endeavour of gathering thought leaders that proceed from diverse cultural and religious traditions could seem a vain effort, perhaps useless. Is it possible to find convergence not only at the level of principles but also when we suggest establishing valid guidelines in a globalized world in which opposing views appear almost impossible to reconcile?

The possibility of meeting scholars and experts from different religious backgrounds became reality in an exciting experience in Jerusalem (2009) and then in Rome (2011). These encounters helped me to understand how necessary and important it is to get to know one another better. The core of our mission in the UNESCO Chair in Bioethics and Human Rights consists in creating a forum of diverse bioethics leaders, delivering a common framework to guide the application of bioethical principles, and informing and influencing ethical, legal and public opinions, decisions and actions relative to medicine, life sciences, and human rights and responsibilities.

For me, these international gatherings are unforgettable and unbelievable. A visible outcome of this recent workshop on the Human Vulnerability is the publication we offer here. Other beautiful experiences are invisible—like all spiritual goods—but not for that reason less real and intense. Among them, I give thanks to God for the opportunity of forging these new friendships.

Alberto Garcia

Contents

Contributors

Dariusch Atighetchi Islamic Bioethics, Faculty of Theology, Lugano, Switzerland

Yechiel Michael Barilan Sackler Faculty of Medicine, Tel Aviv University, Tel Aviv, Israel

Jonathan Keung Lap Chan Centre for Applied Ethics, Hong Kong Baptist University, Hong Kong, People's Republic of China

Prakash Desai The University of Illinois, Chicago, USA

Colleen M. Gallagher Section of Integrated Ethics in Cancer Care, The University of Texas MD Anderson Cancer Center, Houston, Texas, USA

Alberto Garcia UNESCO Chair in Bioethics and Human Rights, Rome, Italy

Adina Halevy Jerusalem, Israel

Jonathan Halevy Shaare Zedek Medical Center, Jerusalem, Israel

Henk ten Have Center for Healthcare Ethics, Duquesne University, Pittsburgh, USA

David Heyd The Hebrew University of Jerusalem, Jerusalem, Israel

Soraj Hongladarom Department of Philosophy and Center for Ethics of Science and Technology, Chulalongkorn University, Bangkok, Thailand

Ping Cheung Lo Centre for Applied Ethics, Hong Kong Baptist University, Hong Kong, People's Republic of China

Martha Tarasco Michel Faculty of Bioethics, Anáhuac University, Mexico City, Mexico

Gonzalo Miranda School of Bioethics, Ateneo Pontificio Regina Apostolorum Rome, Italy

Vasantha Muthuswamy Indian Council of Medical Research, New Delhi, India

Mustafa Abu Sway Integral Chair for the Study of Imam Al-Ghazali's Work, Holy Al-Aqsa Mosque and Al-Quds University, Jerusalem, Palestine

Joseph Tham School of Bioethics, Ateneo Pontificio Regina Apostolorum, Rome, Italy

Stamatios Tzitzis Centre national de la recherche scientifique (CNRS), University Panthéon Assas Paris II, Paris, France

H. Ucko World Council of Churches, Geneva, Switzerland

Ellen Y. Zhang Centre for Applied Ethics, Hong Kong Baptist University, Hong Kong, People's Republic of China

About the Authors

Mustafa Abu Sway is currently the Integral Chair for the Study of Imam Al-Ghazali's Work at the Holy Al-Aqsa Mosque and Professor of Philosophy and Islamic Studies at Al-Quds University in Jerusalem. He earned his PhD from Boston College. He taught at the International Islamic University in Malaysia, and was a visiting Fulbright Scholar at Florida Atlantic University, as well as a visiting professor of Islamic Studies at Bard College. His works include: *Islamic Epistemology: The Case of Al-Ghazzali* (1995, 2004), and *The Religious Rulings of Al-Ghazzali* (1996).

Dariusch Atighetchi was born in Italy and he obtained his degree in philosophy at the University of Milan. Subsequently he received a diploma from the Italian Institute for Middle Eastern and Far Eastern Studies (ISMEO) in Iranian language and culture and a Ph.D. in Comparative Law at the Second University of Naples. For a number of years he has been teaching Islamic bioethics at the Second University of Naples and at other Italian and foreign universities. At present, he is Professor of Islamic Bioethics at the Faculty of Theology, Lugano (Switzerland). His writings include *Islam e Bioetica, Islamic Bioethics: Problems and Perspectives,* and *Islam, Musulmani e Bioetica.*

Yechiel Michel Barilan is a practicing physician and an associate professor in Tel Aviv University's School of Medicine. His first book, on human dignity and human rights was published in 2012 by MIT press; his book on Jewish Bioethics has just been published by Cambridge University Press.

Jonathan Keung Lap Chan received his Ph.D. from the Chinese University of Hong Kong. He is currently Associate Professor at the Department of Religion and Philosophy, Faculty of Arts and Associate Director of the Centre for Applied Ethics, Hong Kong Baptist University. He is Editor of the International Journal of Chinese and Comparative Philosophy of Medicine. His research interests include Bioethics, War Ethics, Chinese Philosophy (Confucianism and Daoism), Informal Logic and Comparative Philosophy. He is author of numerous articles on war ethics, bioethics, human rights, and informal logic.

Prakash N. Desai was born and raised in India. He obtained his medical education at the Faculty of Medicine of the M.S. University of Baroda and completed his psychiatric residency at Northwestern University Medical School in Chicago, Illinois. His extensive work experience includes positions at the B. M. Institute in Ahmedabad, academic psychiatry, Community, Social and Cultural psychiatry, Director of Department of Mental Health and Developmental Disabilities of the State of Illinois, Veterans health care facility on the west side of Chicago, professorial appointment at the University of Illinois at Chicago, Director of Education for the University of Illinois Hospital, and associate member of the Committee of South Asian Studies at the University of Chicago. Among his many publications are *Health and Medicine in the Hindu Tradition* and *Triumph and Tragedy: Psychohistorical Decisions of Mahatma Gandhi*, several book chapters, and over a dozen papers in professional journals. He has received numerous awards for his work including the Oscar Pfeister Award, the George Tarjan Award and the Bruce Boyer Award. He was also actively involved in the governance of the American Psychiatric Association. He is married is the father of two daughters and grandfather of three small children.

Colleen M. Gallagher has worked in the fields of social work and health care ethics since 1985. Dr. Gallagher is currently is the Chief and Executive Director of the Section of Integrated Ethics in Cancer Care, and an Associate Professor, in the Department of Critical Care for The University of Texas MD Anderson Cancer Center where she has worked since 2008. Prior to this she served as System Director of Clinical Ethics for Resurrection Health Care in Chicago, Illinois and before this worked for the Catholic Diocese of Cleveland and Catholic Charities Health and Human Services in Ohio. Throughout her career she has worked as a social worker and program director for programs serving the mentally ill and their families, persons with HIV/AIDS, and the bereaved and other vulnerable populations. She received her Doctoral degree in Bioethics in Health Care Professions and Social Work from The Union Institute and University, Cincinnati, Ohio in 2004. She completed an ethics fellowship in 1997 at Georgetown University, Washington DC. She has also served in assisting the growth of both the social work and bioethics fields having held several positions in professional organizations.

Alberto García is Doctor in Law from Complutense University in Madrid. Presently, he is Professor of Philosophy of Law and International Law at the School of Bioethics of Athenaeum Pontificium Regina Apostolorum in Rome. Furthermore he is a researcher of the Human Rights Institute at Complutense University. For 5 years he has been member of the Steering Committee of Bioethics in the Council of Europe and in 2005 he has been honoured with the National Prize of the Spanish Royal Academy of Doctors in the field of legal and social sciences. He is currently the director of the recently formed UNESCO Chair in Bioethics and Human Rights.

Jonathan Halevy is a physician and Associate Clinical Professor of Internal Medicine at the Hebrew University Medical School in Jerusalem, Israel and Director General of Shaare Zedek Medical Center in Jerusalem.

Adina Halevy has a degree of Masters in Social Work and is a Psychotherapist in private practice and a consultant to ICELP (The International Center for the Enhancement of Learning Potential) in Jerusalem.

David Heyd is Chaim Perelman Professor of Philosophy at the Hebrew University of Jerusalem. His main fields of interest are political philosophy, ethics and bioethics. He is the author of *Supererogation* (Cambridge, 1982), *Genethics* (Berkeley, 1992), *Medical Ethics* (Tel Aviv, 1990) and the editor of *Toleration* (Princeton, 1996). Among his numerous articles, many are concerned with bioethical issues such as procreation, genetic technologies, euthanasia, sex selection, wrongful life, surrogacy and experimentation on human subjects. He served as a member of many ethical committees on surrogacy, on end-of-life issues, on ethics and genetics and the university's Helsinki committee.

Soraj Hongladarom is an associate professor of philosophy and Director of the Center for Ethics of Science and Technology at Chulalongkorn University in Bangkok, Thailand. He has published books and articles on such diverse issues as bioethics, computer ethics, and the roles that science and technology play in the culture of developing countries. His concern is mainly on how science and technology can be integrated into the life-world of the people in the so-called Third World countries, and what kind of ethical considerations can be obtained from such relation. A large part of this question concerns how information technology is integrated in the life-world of the Thai people, and especially how such integration is expressed in the use of information technology in education. He is the co-editor of *Information Technology Ethics: Cultural Perspectives*. His works have also appeared in *Bioethics, The Information Society, AI & Society, Philosophy in the Contemporary World*, and *Social Epistemology*.

Ping-cheung Lo has two earned doctorates (Yale, Religious Studies, 1990; SUNY-Buffalo, Philosophy, 1982). He has been teaching at Hong Kong Baptist University since 1990, and is Professor in the Department of Religion and Philosophy, and Director of the Centre for Applied Ethics. His research areas are: comparative religious ethics, comparative bioethics, and comparative warfare ethics.

Gonzalo Miranda is currently the Dean of the School of Bioethics, Regina Apostolorum Pontifical university. He serves as member of the Steering Committee of the Center for Bioethics of the Catholic University of Sacred Heart, Rome and Board Member and Treasurer of the International Federation of Centres and Institutes of Bioethics of a Personalist Inspiration (FIBIP). He is also member of the Board of Directors of the journals *Medicine e Morale* (Rome), *Medicina y Ética* (Mexico) and *Vida y Ética* (Argentina) and chief editor of *Studia Bioethica* (Rome). He is Professor of Bioethics in the School of Bioethics of Regina Apostolorum and the Catholic University of Sacred Heart, Rome. He is author and editor of and contributed to several books on bioethics, including: *The Stem Cell Dilemma. For the Good of All Human Beings?* (2002); *Bioética e eutanásia* (2000); *Risposta d'amore. Manuale di teologia morale fondamentale*, (2001); *Le radici della Bioetica* (1998); *"Cultura della morte": analisi di un concetto e di un dramma* (1997).

Vasantha Muthuswamy is a physician and retired as Senior Deputy Director General and Chief of Division of Basic Medical Sciences, Traditional medicine & Bioethics and Division of Reproductive Health & Nutrition from Indian Council of Medical Research (ICMR) New Delhi. She was a WHO Fellow at the Kennedy Institute for Ethics, Georgetown University. She is well recognised for bringing out the ICMR's "Ethical guidelines for biomedical research on human subjects" in 2000 and the revised version "Ethical guidelines for research on human participants" in 2006. She was also responsible for the Guidance document for Animal experimentation, Guidelines for Stem cell research and therapy, Guidelines for Safety evaluation of food derived from GE plants, Guidelines for Good Clinical laboratory practices and Guidelines for Safety evaluation of Probiotics. She was also involved in development of guidelines by WHO TDR and UNAIDS and national guidelines on research ethics at Nepal, Sri Lanka and Maldives. As the Founder Secretary of Forum for Ethics Review Committees in Asia Pacific (FERCAP) she was a faculty member in Bioethics workshops/conferences in more than 30 countries. She is currently President of FERCI (Forum for Ethics Review Committees in India) and Chairperson of the Forum for Medical Ethics Society (FMES). She has received Lifetime achievement Award from Indian Society for Clinical Research (ISCR), National Bioethics Conference (NBC) and FERCAP.

Martha Tarasco received her medical degree from the Anahuac University in Mexico and obtained medical specialization in Phoniatry in the National Institute of Human Communication of Mexico. She later received training in the UNAM Program for Clinical Medical Resident Researchers and sub-specialization in Phoniatry in the University of Salamanca in Spain. Since 1986, she has worked privately in her medical office. She has worked in the Medical School of the Anahuac University coordinating the clinical cycles of students. In 1991, she organized a multi-religious symposium on abortion for medical students. During 1992, she was trained in the Istituto di Bioetica of the Universitá del Sacro Cuore in Rome. In 1993, she opened the first Master Degree of Bioethics in Mexico. In 2002 was the first executive director of the newly founded School of Bioethics at the Anahuac University, where she now works as a professor and researcher, and leads the Institute of Humanism in Health Sciences. In 2005 she obtained her Doctoral degree with summa cum laude at the University of Santiago de Compostela, Spain. She has published 8 books, 27 chapters in different books and 37 articles in different journals.

Henk ten Have studied medicine and philosophy at Leiden University, the Netherlands. He received his medical degree in 1976 from Leiden University and his philosophy degree in 1983. He worked as a researcher in the Pathology Laboratory, University of Leiden (1976–1977), as a practising physician in the Municipal Health Services, City of Rotterdam (1978–1979), and as a Professor of Philosophy in the Faculty of Medicine and Faculty of Health Sciences, University of Limburg, Maastricht (1982–1991). From 1991 he was a Professor of Medical Ethics and the Director of the Department of Ethics, Philosophy and History of Medicine in the University Medical Centre Nijmegen, the Netherlands. In September 2003 he joined UNESCO as Director of the Division of Ethics of Science and Technology. Since July 2010 he is Director of the Center for Healthcare Ethics at Duquesne University in Pittsburgh, USA.

Joseph Tham was born in Hong Kong and immigrated to Canada at the age of fifteen. At the University of Toronto, he first majored in Mathematical Sciences and then graduated from Medical School. After several years of work as a family physician, he entered the seminary of the Legionaries of Christ and was ordained a priest in 2004. As a part of this preparation, he has obtained his degrees in philosophy and theology at Rome's *Regina Apostolorum* Pontifical university, where he also completed his post-graduate studies in bioethics. He successfully defended his doctoral dissertation with high honors on "The Secularization of Bioethics—A Critical History" under the direction of Dr. Edmund Pellegrino, former Chairman of the President's Council on Bioethics. He presently teaches bioethics in *Regina Apostolorum*. He is the author of *The Missing Cornerstone* (2004, 2009), *The Secularization of Bioethics* (2007) and *Bioetica al Futuro* (2010). He is the editorial coordinator of *Studia Bioethica* and a Fellow of the UNESCO Chair in Bioethics and Human Rights.

Stamatios Tzitzis has received degrees of Doctor in Law, Master in Political Sociology, Master of Philosophy of Law, Master of Political Science, and LLM. He is currently Research Director of CNRS, Deputy director of the Paris Institute of Criminology at the Panthéon-Assas University (Paris II), President of the International Team of Interdisciplinary Criminal Philosophy at the Paris Institute of Criminology, Panthéon-Assas University and Associate Professor at Laurentian University (Canada). He has published in French, Italian, Portuguese, and Greek, including: *Qu'est-ce que la Personne*? (1996); *Filosofia Penale*, (1999); *La Vittima et il Carneficie* (2004); *Il Diritto e il Valore dell'Uomo*, (2010); *Introduction à la Philosophie du Droit*, (2011). His main areas of interest are philosophy of criminal law, philosophy and bioethics.

Hans Ucko is from Sweden and of Jewish origin. He became in his early twenties a Christian and was after theological studies in Lund, Sweden ordained in the Church of Sweden, which from the beginning used him in various capacities relating to Jewish-Christian as well as interreligious dialogue. He holds a D.Th. from The Senate of Serampore in India comparing contextual Asian theological use of the concepts "people" and "peoplehood" in the Old Testament and their equivalent use in the Jewish tradition. He was 1989–2008 the Program Secretary for Interreligious Relations and Dialogue of the World Council of Churches. He has held the position as President of Religions for Peace Europe. Now retired, his interfaith contribution is focused on a project on a social and spiritual contract for a plural Europe as well as a multifaith network on the UN Convention on the Rights of the Child. He is married, has three children and now lives in France.

Ellen Y. Zhang holds a Ph.D. in Philosophy of Religion from Rice University (Houston, TX). She is now an associate professor in the Department of Religion and Philosophy at Hong Kong Baptist University. She is also a research fellow at the Center for Applied Ethics at HKBU. Her research projects are related to Chinese philosophy, ethics, and comparative studies. Her publications in ethics include "'Weapons Are Nothing but Ominous Instruments': The *Daodejing*'s View on War and Peace" (2012), "The Ethic of Givenness: An Ethical Dimension of Jean-Luc Marion's Theology of *Gift* and the Buddhist Principle of *Dana*" (2011), "Community, the Common Good, and Public Healthcare: Confucianism and its Relevance to Contemporary China" (2010), "What Is Personhood? Kant and Huayan Buddhism" (2010), "The Unity of Corporeality and Spirituality: Love in Daoism" (2012), "The Neo-Confucian Concept of Body and Its Ethical sensibilities" (1999). She is now the editor-in-chief for *International Journal of Chinese and Comparative Medicine*. She is the member of the APA (American Philosophical Association), the AAR (American Academe of Religion) and the APCA (Association of Chinese Philosophers in America).

Chapter 1
Introduction: The Principle of Vulnerability: Meeting Ground of Six Religions

Joseph Tham

Anyone familiar with the history of bioethics will agree that it emerged as a result of public scandals of unethical experimentations on vulnerable groups of the American society. One can recall the famous 1966 whistle-blowing article by Dr. Henry Beecher in the *New England Journal of Medicine*, the Tuskegee and the syphilis scandals as landmark events in the nascent field of bioethics.[1]

The protection of vulnerable populations in biomedicine has since then become indispensible in any bioethical consideration. The recent news of experiments done in Guatemala only highlights this perennial temptation to exploit the weak for the sake of scientific knowledge and progress.[2] But one does not have to look very far into the past to see that the most vulnerable need protection. For instance, two neurosurgeons from a reputable US university hospital were just found guilty of infecting brain tumor cells of dying patients as a form of treatment.[3]

Hence, it is not surprising that protection of human vulnerability and personal integrity became one of the articles promulgated in the 2005 UNESCO Declaration on Bioethics and Human Rights.[4] Article 8 of this declaration states:

> In applying and advancing scientific knowledge, medical practice and associated technologies, human vulnerability should be taken into account. Individuals and groups of special vulnerability should be protected and the personal integrity of such individuals respected.

Yet, vulnerability is a latecomer to the scene of bioethics already dominated by the language of justice, autonomy, beneficence, dignity, rights, and virtues. In fact, after this UNESCO declaration, some bioethicists wonder if there is really a need for this new principle in bioethics. The article written by Henk ten Have, the then

[1] Jonsen (1998); Rothman (1991).

[2] Beecher (1966).

[3] Lundstrom (2012).

[4] http://www.unesco.org/new/en/social-and-human-sciences/themes/bioethics/bioethics-and-human-rights. Accessed 28 August 2013.

J. Tham (✉)
School of Bioethics, Ateneo Pontificio Regina Apostolorum, Rome, Italy
e-mail: jtham@legionaries.org

J. Tham et al. (eds.), *Religious Perspectives on Human Vulnerability in Bioethics*, Advancing Global Bioethics 2, DOI 10.1007/978-94-017-8736-9_1,

acting director of UNESCO's Division of Ethics of Science and Technology who overlooked the drafting of the declaration, and the account of Fr. Gonzalo Miranda who represented the Holy See as a participant of this process are illuminating in this regard. In response to these critiques, the International Bioethics Committee as a UNESCO working group has dedicated the last several years to a better understanding of this principle of human vulnerability and personal integrity. It made its final report generated in 2011.[5]

By its very nature, United Nations documents tend to be non-religious and non-sectarian. In this volume, however, we wish to supplement the analysis of vulnerability from different religious perspectives. The genesis of this effort is due to a convergence of three different strands from the editors of this book. Fr. Miranda, as noted, participated as an observer during the drafting UNESCO Declaration. Among the long process of drafting and debate, there was only one session where religious representatives were asked to give their views. Fr. Miranda spoke on behalf of the Christian-Catholic tradition, whereas other speakers spoke from Islamic, Jewish, Buddhist and Hindu perspectives. At the end of this session, the moderator of the panel lamented that it is a pity UN organizations do not pay more attention to religion since a great majority of the world's population is affiliated with some religious traditions.

I myself became interested in the role of religion in bioethics after my doctoral dissertation on the subject of "The Secularization of Bioethics" which is a historical analysis of how religion and theological input became marginalized in American bioethics for the last 40 years (Tham 2008). There has been a general rejection of religion in this field because it is believed that religion can cause tension as it is not inclusive enough. As a consequence of this research, I became engrossed in the question of the place of religion in bioethics, and how religion might supplement or correct the overly secular tendency of the discipline (Tham 2009).

The third contribution comes from Alberto Garcia, who as a law professor has participated as consultant in the areas of bioethics in the European Commission. When he arrived at Rome to teach at the School of Bioethics, where both Fr. Miranda and I were professors, he initiated the project to establish a *UNESCO Chair on Bioethics and Human Rights*. In 2009, UNESCO signed an agreement to establish this Chair with two Roman Universities, the European University of Rome and the Pontifical Athenaeum Regina Apostolorum. The UNESCO Chair wishes to offer a university framework of reflection and study, providing information and fostering the application of bioethical principles in science, medicine and new technologies based on the aforementioned *Declaration on Bioethics and Human Rights*. Not being aware of the others' interests, one of the areas Alberto proposed for the Chair to dedicate its research was "Bioethics, Multiculturalism and Religion."

[5] Final report of 2011 http://unesdoc.unesco.org/images/0018/001895/189591e.pdf, and earlier drafts of 2010 http://unesdoc.unesco.org/images/0018/001891/189194e.pdf. Accessed 28 August 2013. http://unesdoc.unesco.org/images/0018/001895/189591e.pdf. Accessed 28 August 2013 http://unesdoc.unesco.org/images/0018/001895/189194e.pdf; and 2009. Accessed 28 August 2013 http://unesdoc.unesco.org/images/0018/001855/185533e.pdf. Accessed 28 August 2013.

Under this auspice, experts in bioethics coming from the different religious traditions were brought together to discuss and cooperate on the pressing ethical questions facing humanity in a rapidly advancing medical, legal and technological world. The idea was to establish a permanent academic forum to promote dialogue and bioethical reflection in the light of human rights and duties addressed from different religious and cultural perspectives. The first attempt gathered Jewish, Christian, and Islamic scholars in Jerusalem from December 13–14, 2009.[6] The format of this event was mostly conferences with some discussion and dialogue. The conclusions were summarized and a general understanding was reached and formalized in a two-page document.[7]

When Fr. Miranda, Alberto and I analyzed our first experience and discussed follow up activities, we decided that a workshop format would be most appropriate for the next gathering.[8] We soon agreed that the topic of our second meeting would discuss article 8 of the UNESCO Declaration, since it was evident that while each religious tradition is different, they all carry within themselves the mission to protect the weak, the underprivileged, and the poor.

Hence, in our second international gathering that took place in Rome from October 9–11, 2011, we focused on what vulnerability means in bioethics from different religious and cultural perspectives. We strived this time for greater diversity both in terms of variety of religions (by including Buddhism, Confucianism, Hinduism in addition to the monotheistic religions), within each tradition (for example, inviting voices from Protestant and Orthodox Christians in a majority Catholic presence) and in terms of interdisciplinarity and nationality. It was not easy to fill the roster of speakers, especially because of the paucity of bioethical reflections in some of the religious groups. In a way, the speakers chosen were not based on their international fame (even though some speakers were quite notable in the field) but rather from personal contacts and network of friends. This approach was based on the belief that academic gatherings should not only be an intellectual interchange, but also a meeting of hearts which creates communion.

This has helped the workshop to succeed in spite of tense moments that sometimes transpire when strong religious convictions confront each other, complicated by certain geo-political backgrounds of the participants.

At this workshop, 16 scholars hailing from China, France, India, Israel, Italy, Mexico, Palestine, Spain, Switzerland, Thailand and USA presented papers on how their religious and cultural traditions offer support and protections to the vulnerable in society. At the planning stage, we found it useful to have two sets of papers from each of the six religious groups: the first set relates to the foundational questions of

---

[6] http://www.unescobiochair.org/index.php?option=com_content&view=article&id=59&Itemid=29&lang=en. Accessed 28 August 2013.

[7] http://www.unescobiochair.org/index.php?option=com_content&view=article&id=70%3Aconclusions&catid=50&Itemid=103&lang=en. Accessed 28 August 2013.

[8] The Common Good for the Twenty-First Century: Beyond Individualism and Collectivism—A Sino-American Dialogue (2009, October 29–30,) University of Notre Dame Center for Ethics and Culture and Hong Kong Baptist University Center of Applied Ethics at the Hong Kong Baptist University. http://cae.hkbu.edu.hk/Conferences/091029_conference.html. Accessed 28 August 2013.

vulnerability from religious perspectives, and the second relates to attitudes, obligations or treatments towards the vulnerable in these traditions. For the first set, each representative was commissioned to write papers on: (a) the anthropological basis of vulnerability (e.g., questions on human contingency, suffering and mortality); (b) the ethics of vulnerability (e.g., the attitudes of respect, compassion, mercy, charity or love); and c) the legal aspects of vulnerability (both religious and civil).

For the second set of papers, we invited each of the six religious groups to write a paper addressing certain individuals who are especially vulnerable and how they are perceived and cared for in their respective religious traditions: (a) Children before and after birth; (b) Women; (c) Physically and mentally handicapped; and (d) The elderly. These four categories of vulnerable groups came from the 2010 IBC working document on the Principle of Vulnerability.[9]

After the scholars from the six religions submitted the papers, they were made available for the others at the workshop to study before they come together. Having the chance to read each other's papers allowed the workshop session to proceed expediently. After the authors gave a brief summary of their papers, they were given ample time to discuss, clarify, question, challenge and criticize each other's thoughts. The rich interchange among the presenters and from the audience at the workshop provided a better understanding of each religious tradition regarding their understanding of vulnerability. These papers were eventually modified based on these interchanges and they are found here in this present form.

It was clear from the beginning that with so many different religious traditions, and traditions within each tradition, many conceptual problems with vulnerability were raised. The East-West contrast made discussion very interesting, and at times heated. For instance, according to Buddhism, vulnerability is not something negative, but part of the human condition that we need to accept as a fact of life. There are also certain trends that can be detected across the board. A recognition of our human limitations and a need for response is present in all religions, albeit with different names: Mercy (*hesed*) in Judaism, Agape for Christians, Humaneness (*ren*) in Confucianism, and Compassion in Buddhism, etc.

At the same time, many participants of the workshop find the formulation of vulnerability based on the human rights language too individualistic. This is somewhat foreign to major religions where the self does not exist in isolation, but is normally immersed in a web of relations—family, friends, religious community, and society. The emphasis on those who are vulnerable demanding their rights to be protected or given special attention is critiqued as predominantly a Western liberal value, which in bioethics is translated to mean patient autonomy, self-determination and free choice. In contrast, Confucians place more emphasis on family decisions. This is echoed in different ways in Hinduism, Islam and Judaism which speaks more

[9] http://unesdoc.unesco.org/images/0018/001891/189194e.pdf. Accessed 28 August 2013. It is interesting to note that in the 2011 final report, these categories of vulnerable groups disappeared, probably due to the fact that by labeling certain groups as vulnerable, they become stigmatized. This issue also became evident in the workshop itself when several papers emphasized the fact that everybody is vulnerable to a certain extent, and singling out particular groups might actually be counterproductive.

of duties than rights towards the weak and underprivileged. All three branches of Christianity represented at the workshop were not totally comfortable with the liberal exaltation of subjective individualism.

This is not to say that individual rights are unimportant. In today's democratic societies, laws have been drafted to protect individuals and communities against slavery, discrimination, torture or genocide. Yet, there are enormous challenges for these age-old cultures as they catch up with modernity. For instance, India needed to forego the caste system as unjust and discriminatory, even though this has been ingrained in millennium-long tradition of Hinduism. Islamic traditions of laws and customs derived from Koranic revelation is at time at odds with international standards on issues such as the status of women or female circumcision. Christianity has been confronting modernity for the last few centuries in the process called secularization. While there are definite traces of Judeo-Christian influence in the genesis of human rights, there is palpable unease to include abortion and same-sex marriage as part of these rights.

Any attempt to compare different cultural and religious traditions on a particular topic is bound to face the question of ethical relativism on the one hand and *incommensurability* of ethical traditions on the other. In other words, is it realistic to expect convergence of ideas coming from such a wide spectrum of backgrounds and positions? Are these ethical and religious traditions comparable to one another, or are they incommensurable? Since differences are inevitable, is it possible to judge the superiority of one tradition over another? If not, are we left with a sort of ethical relativism? Can any type of universal ethics resolve incompatible views on a particular bioethical issue due to cultural differences? There are no simple and easy answers to these questions.

This issue has been amply analyzed by philosopher Alasdair MacIntyre who approaches the question of moral inquiry as "tradition-constituted" in *Whose Justice? Which Rationality?* According to this fascinating work, one cannot be an independent observer beyond a particular tradition speaking to all parties, but can only inquire from within a particular moral tradition to which one belongs. Macintyre further applies this paradigm in an article entitled "Incommensurability, Truth and the Conversation between Confucians and Aristotelians about the Virtues."

These two virtue-based traditions are compared precisely because the many commonalities mistakenly lead many scholars to gloss over their incommensurable differences. For example, he cites a strong interdependence among Aristotelian virtues which is absent in the Confucian view; and the Confucian prerequisite of performing exterior rituals (*li*) in the practice of right action (*yi*) would not make sense for Aristotle or Aquinas.

Yet for MacIntyre this incommensurability does not lead to relativism, "Incommensurability, it turns out, does not preclude rational debate and encounter" (MacIntyre 1991).

Likewise, he dismisses the claim that incommensurability is merely a problem of translation, even though many concepts and terms do depend on a cultural milieu.[10]

[10] MacIntyre claims on the contrary that, "Incommensurability may, but need not, be associated with and arise from untranslatability" (MacIntyre 1991). See too Davidson pp. 183–198.

To enter the conversation, what is required—other than being sufficiently fluent in the languages of both traditions—is that the inquirer must be fully immersed in his or her own culture and history in order to accurately represent it. The first stage requires scholars to write a critical account of their own tradition, the development and history of its theory and practice, its successes and failures, challenges and crises. Rational encounters with rival civilizations take place in every authentic tradition throughout history, when coherence of customs and ethos are measured and tested, resulting in processes of adaptation, absorption and purification. (Nie and Campell) The second stage involves the more serious task when inquirers of one moral tradition write the history of the rival moral tradition from that rival tradition's point of view, "employing the standards of rational success or failure internal to that other's point of view."[11]

Two conditions are necessary for this to happen. One must be prepared to expose one's own tradition with intellectual honesty and "maximal vulnerability" without hiding defects. Second, one must recognize that there is not a neutral, independent standpoint to judge between rival traditions. In this sincere conversation, such exchanges and comparisons would allow the rival traditions to see their weaknesses and strengths, and rationally recognize their own incoherence and the superiority of their rival, with the possibility of abandonment of their own tradition. In a similar vein, the Catholic encyclical *Caritas in Veritate* also encourages in interreligious dialogue to promote peaceful co-existence and solidarity, while at the same time avoiding the danger of cultural relativism or eclecticism (Benedict XVI).

The inclusion of religious and intercultural dimension in bioethical discussion is of great and relevant dimension today, where religion is sometimes seen as a cause of division rather than peace. In this regard, the basis of intercultural dialogue is based on the belief that reason is common to all humanity of any faith or the absence of it. When reason itself is not closed to religious input, it can then dialogue with ease with bioethics from other religious traditions. This engagement is possible when reason is open to faith, while faith-based assumptions are also open to the critique of reason, thus faith and reason purify each other from possible excesses (Ratzinger and Habermas). Fortunately in bioethics, there are substantial agreements on many issues, for example the rejection of human or sex trafficking, using humans as products or body parts, female genital mutilation, etc. Other issues must still be debated. For this reason, human rights and human dignity language can be useful in an international setting with certain legal force, on the condition that it restrains itself from excessive liberal extensions of rights, that it reconsider its link to natural rights, and not *a priori* ostracize religion from discussions.

As one can imagine, tradition-constituted conversation with well-versed religious scholars can be pain-staking, slow and arduous. We believe that this book

[11] "Aristotelians need to understand the history of Confucianism as a form of moral inquiry and practice, as it has been, is or would be written from a Confucian point of view, in order to be able to learn to identify those epodes in which Confucianism becomes in some way problematic for a sufficiently tough minded and insightful Confucian." And vice versa for Confucians to understand Aristotelians." (MacIntyre 1991).

resulting from such a rare gathering of experts is a first step in this direction. This first step consisted in opening a space where an atmosphere of friendship and mutual esteem reigns. This will permit sincere dialogue and encounters and allow us to see the other as our brothers and sisters in our common humanity. In this space, religious traditions can know and confront each another in a spirit of humility and openness to the truth, without ideologies or political agendas that often color such encounters.

We hope and pray that as a result of this humble effort, we can eliminate suspicions that are sometimes the cause of distrust and even violence in our globalized world.

Bibliography

Beecher, H. K. 1966. Ethics and clinical research. *New England Journal ofMedicine* 274:1354–1360.

Cf. Benedict, XVI. 2009. Encyclical Letter Caritas in Veritate http://www.vatican.va/holy_father/benedict_xvi/encyclicals/documents/hf_ben-xvi_enc_20090629_caritas-inveritate_en.html

Cf. Davidson, D. 1984. On the very idea of a conceptual scheme. In *Inquiries into truth and interpretation,* eds. D. Davidson. Oxford: Clarendon.

Cf. MacIntyre, A. 1988. *Whose justice? Which rationality?* Notre Dame: University of Notre Dame Press.

Cf. Nie, J. B., and A. Campbell. 2007. Multiculturalism and Asian bioethics: Cultural war or creative dialogue? *Journal of Bioethical Enquiry* 4 (3): 163–167.

Cf. Ratzinger, J., and J. Habermas. 2007. *The dialectic of secularization: On reason and religion.* San Francisco: Ignatius.

Jonsen, A. 1998. *The birth of bioethics*. Oxford: Oxford University Press.

Lundstrom, M. 2012. 2 UC Davis neurosurgeons accused of experimental surgery are banned from human research. The Sacramento Bee http://www.sacbee.com/2012/07/22/4648415/2-uc-davis-neurosurgeons-accused.html. Accessed 28 Aug 2013.

MacIntyre, A. 1991. *Incommensurability, truth and the conversion between confucians and aristotelians about the virtues culture and modernity: East-West philosophical perspectives*. Honolulu: University of Hawaii.

Rothman, D. 1991. *Strangers at the bedside: A history of how law and bioethics transformed medical decision making*. New York: Basic Books.

Part I
General Considerations on the Principle of Vulnerability in Bioethics

Chapter 2
Vulnerability: How did the principle Come About?

Gonzalo Miranda

2.1 Introduction

In June 2005, delegates of the Member States of UNESCO gathered in Paris to review the draft of the *Universal Declaration on Bioethics and Human Rights*[1]. I was present as a delegate of the Holy See. Immediately after the approval of the "Principle of Vulnerability,"[2] one of the delegates, a friend, came to me and said: "It's a miracle! Do you realize the implications? Who is the most vulnerable among the vulnerable?" "Yes, I know, I know," I replied. The implicit reference was to unborn human beings: embryos and fetuses.

This little anecdote helps us see why it may be interesting to examine the genesis of the Principle of Vulnerability during the drafting of the Declaration.

2.2 A New Principle

The text of the Principle, in article 8, states: "In applying and advancing scientific knowledge, medical practice, and associated technologies, human vulnerability should be taken into account. Individuals and groups of special vulnerability should be protected and the personal integrity of such individuals respected."

It is true that—as stated at the beginning of the Report on the principle recently published by of the International Bioethics Committee of the UNESCDO (IBC)—"This notion is not new. The concept of vulnerability appears in important national documents, starting with the US Belmont Report of 1978, and in international docu-

[1] From now on "the Declaration."

[2] From now on "the Principle."

G. Miranda (✉)
School of Bioethics, Ateneo Pontificio Regina Apostolorum, Rome, Italy
e-mail: g.miranda@arcol.org

J. Tham et al. (eds.), *Religious Perspectives on Human Vulnerability in Bioethics*, Advancing Global Bioethics 2, DOI 10.1007/978-94-017-8736-9_2,

ments, such as the third and most complete version of the International Ethical Guidelines for Biomedical Research Involving Human Subjects of the Council for the International Organizations of Medical Sciences (CIOMS) (2002) and in the latest (2008) version of the Declaration of Helsinki, which makes specific reference to vulnerability in articles 9 and 17" (IBC 2011).

Nevertheless, the concept of vulnerability was not really present in the texts of Bioethics or in the official documents related to medical ethics. Indeed, in the original draft of Declaration prepared by the IBC, there was no article stating the Principle of Vulnerability.

Let us see, therefore, how and under what circumstances, the Principle arose.

2.3 The Difficult Process of Revision of the Declaration

This Principle, as I said, did not exist in the text prepared by the ICB and proposed to the delegations of the Member States of UNESCO for revision. It was during this review process, and specifically in the meeting of June 2005, that the principle was introduced in the Declaration. That in itself is interesting, since the fundamental task of the delegates was not to create new principles of bioethics, but more simply to review the proposals made by the IBC.

It should be noted that the deliberations of the delegates, and the revision of the text by all of us, was not an easy or a simple task at all. Let me give an example.

In the previous session to that of June, the review process had reached a stalemate, because of the frontal opposition among the delegates who demanded the introduction of an expression for the respect for human life and those who absolutely refused it. One of the delegates opposed to it told me very clearly that his government could not accept that statement, because in his country "therapeutic cloning" is allowed. It took three days, during the meeting in June, to arrive at a consensus on this point (while advancing in parallel on several articles of the Declaration). Some delegates had to ask for permission from their governments before they finally agreed to accept the wording: it would not be a principle, but only be mentioned in article 2, which sets out the aims of the Declaration. It was decided to add the expression "by ensuring respect for the life of human beings" (IBC 2013).

2.4 Immediate Approval of the Principle

Knowing the complexity of the deliberations in the revision of the Declaration, one can understand our surprise when the proposal of the new Principle of Vulnerability was approved in few minutes.

There had been much work before this on the principle of "Informed Consent." In addition to the complexity and difficulty of the application of this principle, it was evident that most of the delegates were concerned about protecting people who are not able to give true consent in the practice of medicine and in medical research. There was already, in article 5 on autonomy and individual responsibility, an indication of the duty to take special measures to protect the rights and interests of people who are not able to exercise their autonomy. In addition, article 6 had several phrases aimed at protecting individuals and groups most vulnerable to abuse. So many clauses and specifications were added to this section that it was finally decided to split the article into two different items (the current articles 6 and 7).

After finishing this review, one of the representatives asked the floor to propose the introduction of a new article. She was asked to dictate the text, and no one opposed her. The chairman proposed to analyze the new article after the mid-morning break, inviting anyone who wished to do so to work with the delegate who had just made the proposal on improving the draft text. After the break, and after just a few brief touches, the text was approved with unusual speed.

I believe that this phenomenon was due, in part, to the "ethical concerns" which had already been expressed in the review of previous articles. There was a general agreement on the need to protect and respect the weakest, the most vulnerable. Perhaps the rapid approval was also due to a certain weariness among the delegates. It is also likely that some delegates were not really aware of the important implications of those few lines just added to the Declaration.

2.5 Towards a Report on the Principle

As you know, the IBC has been given the task, once the declaration was formally approved in October 2005, to develop and deepen some of the principles shaped in it. The Committee immediately began working on the elaboration of a report on the Principle of Informed Consent and on the one called the Principle of Social Responsibility. During the 14th session of IBC, held in Nairobi in May 2007, a report on the Principle of Vulnerability was also proposed. The proposal was justified based on the novelty of the principle, and because of its priority over other principles of the Declaration, since it describes a general human condition. Some of the problems arising from developing the Principle were also highlighted.

The discussion that followed, by members of the committee and some observers, seem to me very significant. Some stressed the importance and expediency of the Principle. Others, by contrast, showed a certain "uneasiness" before the Principle and of some of its possible implications.

One member of the committee had suggested earlier—when analyzing the draft of the Report on informed consent—that the protection of those who cannot give consent should not begin from the moment of birth, but should also include the prenatal period, as during gestation harmful actions can be performed against a human being who cannot give consent. This same delegate, commenting on the proposal to study the Principle of Vulnerability, pointed out that it should also be applied to embryos and human fetuses who are also vulnerable. In opposition to this, one observer opined that the concept of vulnerability can only be applied to humans who are conscious, arguing that being vulnerable means to be aware of the possibility of being damaged.

At this point I ventured to ask to speak, in order to remind everyone that according to the Oxford English Dictionary, "vulnerable" simply means: "exposed to the possibility of being attacked or harmed, either physically or emotionally." Therefore, the condition of vulnerability has nothing to do with the eventual consciousness of the individual who can suffer an attack; indeed, individuals who are not aware of the danger are probably more vulnerable, because they cannot defend themselves from it.

Several speakers in public, and some in private in the corridors, proposed that any further study of the Principle of Vulnerability be done not in an *ad-hoc* document but simply as part of the Report on the Principle of Social Responsibility. Some wanted to include it in the Principle of the Protection of Plants and Animals, an enlargement that seemed inadequate to me. It is true that the concept of vulnerability applies perfectly to any reality that can suffer damage: a person, an animal, a plant, and also a statue… But article 8 of the Declaration establishing the Principle speaks of "*human* vulnerability"; and thus the expression: "especially vulnerable individuals and groups" refers only to human beings.

All these seemed to me strategies in order to "water down" the Principle.

Eventually, the idea was approved to develop a report on the Principle of Vulnerability. The Report was published in June 2011.

And today we are here thinking about this Principle, which I consider an important part of the Declaration and of Bioethics and Human Rights in general. Our religious traditions can and should explore the concept—considering its important implications in the medical health sciences—with the sincere desire to promote respect for all, especially the most vulnerable individuals and groups.

References

IBC. 22 June 2011. Report of IBC on the principle of respect for human vulnerability and personal integrity, art. 2, Paris. http://unesdoc.unesco.org/images/0018/001895/189591e.pdf. Accessed 28 Aug 2013.

IBC. 2013. Universal Declaration on Bioethics and Human Rights, art. 2, b. http://portal.unesco.org/en/ev.php-URL_ID=31058&URL_DO=DO_TOPIC&URL_SECTION=01.html. Accessed 28 Aug 2013.

Chapter 3
The Principle of Vulnerability in the UNESCO Declaration on Bioethics and Human Rights

Henk ten Have

3.1 Introduction

"Vulnerability" is a relatively new concept in contemporary global bioethics. In authoritative bioethical documents it first surfaced in the Belmont Report in 1979, as a special consideration in the application of the general principles of respect for persons, beneficence, and justice. This consideration is especially relevant in the context of research with human beings.

The Council for International Organizations of Medical Sciences (CIOMS) refers to vulnerability in its first guidelines for research in 1982. The Declaration of Helsinki starts to refer to vulnerability in its fifth revision of 2000. In both cases, this first use is brief and only presented as a consideration to be taken into account. The notion of vulnerability itself is not explained; also, the criteria for the identification of individuals or groups as vulnerable are not explicitly presented. The status of the notion of vulnerability is changing in the 1990s.

CIOMS Guidelines in 1991 specifically mention vulnerability as a fundamental principle but included in the principle of respect for persons. This position is not reiterated in later Guidelines that present vulnerability as a special application of the principle of respect for persons and the principle of justice. Vulnerability, however, is now upgraded as a guideline for research itself. The UNESCO Declaration on Bioethics and Human Rights, adopted in 2005, is the first international document that articulates respect for vulnerability as an ethical principle. At the same time, the scope of vulnerability is broadened: it is no longer only relevant for medical research but for healthcare as such.

The increasing number of references to vulnerability in global bioethics documents does not imply that it is clear what vulnerability really is and how it should be applied in bioethical discourse. A brief review of authoritative bioethics documents will show that major controversies exist concerning the status of the notion of vulnerability, its content and scope, as well as its consequences for healthcare practice.

H. ten Have (✉)
Center for Healthcare Ethics, Duquesne University, Pittsburgh, USA
e-mail: tenhaveh@duq.edu

J. Tham et al. (eds.), *Religious Perspectives on Human Vulnerability in Bioethics*, Advancing Global Bioethics 2, DOI 10.1007/978-94-017-8736-9_3,

3.2 The Emergence of Vulnerability in Authoritative Bioethics Documents

3.2.1 The Belmont Report

The first reference in the bioethics literature to the notion of vulnerability is in the Belmont Report in 1979. An analytical framework of three principles is presented: respect for persons, beneficence, and justice, as well as three applications to the conduct of research. Vulnerability is mentioned under each of the three applications. In the section on informed consent, it is argued that the element of voluntariness of the consent process can be compromised through undue influence. This may occur for two reasons: either the offer is difficult to resist because the inducement is excessive or because the subject is vulnerable ("…inducements that would ordinarily be acceptable may become undue influences if the subject is especially vulnerable," Belmont Report 1979, p. 23195). The second reference to vulnerability is in the section on assessment of risks and benefits.

Here it is argued that there are five considerations to be taken into account in the assessment of the justifiability of research. One of these is the involvement of vulnerable populations ("When vulnerable populations are involved in research, the appropriateness of involving them should itself be demonstrated," ibidem, p. 23196). Although the condition of the particular population involved should be considered in the judgment, no further explanation is offered. A more elaborated reference is in the section on selection of subjects. The report argues that injustice may appear even if individual subjects are selected fairly. Burdens and benefits of research may still not be distributed fairly at social level.

It continues: "One special instance of injustice results from the involvement of vulnerable subjects" (ibidem, p. 23197). Some examples are provided: racial minorities, the economically disadvantaged, the very sick, and the institutionalized. The Report furthermore explains very briefly what the problem is. These subjects are readily available for research but at the same time should be protected because of "their dependent status and their frequently compromised capacity for free consent." It is also mentioned that as a result of their illness or socioeconomic condition, these subjects are "easy to manipulate" (ibidem, p. 23197).

Vulnerability therefore is mentioned in connection to all three basic ethical principles proposed in the Belmont Report. The references, however, are very brief; they serve to call attention to specific conditions and considerations for the application of the principles. The two volumes of background papers prepared to assist the Commission do provide somewhat more clarity Belmont Report, Appendix 1978. The only person explicitly mentioning vulnerability is Robert Levine. In his paper on the assessment of risk-benefit criteria he uses the term 'vulnerable' to refer to individuals who are most susceptible to harm.

In the same section he refers to populations that are especially vulnerable. As examples he mentions classes of persons with limited capacities to consent (children, fetuses, prisoners, the institutionalized mentally infirm) and also the unconscious and the inebriated (by alcohol, narcotics etc). Another category is persons vulnerable as a consequence of their life situations. An example is that of persons legally capable but in practice incapable to consent. Another example is poor persons and persons with prolonged chronic illness because they are especially motivated to take risks, like other persons (infertile persons, obese persons, depressed persons, persons who believe they are imminently dying) who are desperately seeking a cure or solution and therefore motivated to take more risks. In fact, two different categories of vulnerable persons are distinguished: compromised consent and compromised assessment of risks (Ibidem, pp. 2–53).

However, if we look more carefully, apparently both categories can be explained because the ability to make autonomous decisions is diminished or restricted.

In a special paper on informed consent, within the background materials for the Belmont Report, Levine elaborates on the connection between vulnerability and compromised consent. Individual autonomy can be limited in three cases: the capacity to consent is limited, the individuals have subordinate relationships vis-à-vis the investigator or his institute, or else the individuals have special vulnerability, "by virtue of other aspects of their life situations" (pp. 3–38). Levine gives an example of a vulnerable research population: patients recently admitted to a coronary care unit with known or suspected myocardial infarction. At least some of these patients are vulnerable: they might perceive themselves as in the process of dying, and their abilities to make rational judgments are impaired because they have received a narcotic (pp. 3–50). In this last case, vulnerability refers to 'weakness' but it is still included in the broader category of 'limited autonomy'.

The most extensive discussion of vulnerability is in Levine's paper on the selection of human subjects for research. Levine identifies a broad class of persons not being capable of protecting their own rights and welfare, for three reasons (pp. 4–2):

a. uncomprehending (unable to understand the information necessary to provide valid consent)
b. vulnerable (defined as "those who are either capable of being wounded or defenseless against injury," pp. 4–85)
c. dependent ("unable to exist, sustain himself, or act suitably or normally without the assistance of another or others," pp. 4–90).

Each of these categories consists of a spectrum "ranging from slightly to absolutely incapable," (pp. 4–30). What should be of concern are capabilities that are so reduced that different treatment is required. But in practice the three categories often overlap.

Levine further distinguishes vulnerable individuals in two groups depending on the origin of the vulnerability: the sick and the impoverished, while minorities are a third group of potentially vulnerable (pp. 4–34 ff.). Characteristic of the first group is that the sick role is providing barriers to autonomy. For example, persons with prolonged chronic illness are willing to take any risk for possible relief when they are refractory to standard therapeutic modalities (pp. 4–36). Also certain conditions of disease and illness such as depression, obesity, infertility, and imminent death compromise individual autonomy when people are desperate for cures and assistance. Levine gives relatively most attention to dying persons. The second vulnerable group is the impoverished. This condition is defined as: "a condition in which a person considers it necessary to take extraordinary risks to secure money or other economic benefits that will enable him to purchase what he considers to be the necessities of life" (pp. 4–42). This does not imply that they cannot partake in research. But they are vulnerable in the sense that their circumstances may motivate them to take extraordinary risks. They may, for example, be too poor to purchase medical care. If they have the possibility to participate in research, they will assume that the costs of care will be underwritten by the sponsor of the research. This group will therefore be easy to recruit in research trials so that special precautions need to be taken. For example, they should not be paid to assume risks (only compensated for injury) and economic inducements should not be based on estimated risks but on the amount of inconvenience.

The third vulnerable group is minority groups (determined by race, sex, ethnicity etc). This group is not vulnerable a priori but they have the potential to be vulnerable. They can be the object of discriminatory societal customs that will make them vulnerable in practice. The best protection, according to Levine, is to have this group represented on the Institutional Review Boards.

Characteristic for all three groups is that vulnerability refers to the social and existential conditions that do or can diminish autonomy. As Levine argued, autonomy can also be limited due to lack of comprehension, resulting in compromised consent or compromised assessment of risks; but in this paper he does not consider this as vulnerability (while he did in his paper on the assessment of risk-benefit criteria). This probably was to distinguish internal and external limitations of autonomy. If this distinction is the rationale, then it is unclear why dependency (as another external limitation) is distinguished from vulnerability. As will be shown in the next section, the notion of vulnerability will soon develop into the broader category, subsuming compromised consent, or weakness due to social and existential conditions, and dependency, under the same concept.

3.2.2 The CIOMS Guidelines

The Council for International Organizations of Medical Sciences (CIOMS) has been established in 1949 by WHO and UNESCO. Since the 1980s it is active in the field of bioethics and health policy, issuing international guidelines for biomedical research and epidemiology. In 1982 CIOMS, in collaboration with the World Health Organization, issued *Proposed International Ethical Guidelines for Biomedical Research Involving Human Subjects*. The purpose of these guidelines was twofold. The first relates to the question how the principles of the Declaration of Helsinki (with the first revision adopted in 1975) can be put into practice.

These principles have "universal validity" but their mode of application will vary in different circumstances (CIOMS 1982, p. 222). Different contexts require different applications. This brings in the second purpose: the guidelines focus on research from the perspective of developing countries. At that time biomedical research was only conducted on a limited scale in the developing world, and most countries do not have a national framework for protecting research subjects against abuse. In order to frame the guidelines for research in a way that can be applied in developing countries, CIOMS undertook an extensive consultation process involving experts from 60 developing countries.

The 1982 Proposed Guidelines is the first international document that explicitly refers to vulnerability in a heading in the text. Having underlined that consent of subjects is the first and major issue to consider, the document subsequently elaborates that there are many individuals for whom consent is compromised. Separate sections discuss children, pregnant and nursing women, and the mentally ill and mentally defective persons, followed by the special section 'Other vulnerable social groups': "The quality of the consent of candidate subjects who are junior or subordinate members of a hierarchically structured group requires careful consideration, as willingness to volunteer may be unduly influenced by the expectation, whether justified or not, of adventitious benefits" (CIOMS 1982, p. 223). Examples are medical and nursing students, employees of the pharmaceutical industry, and members of the armed forces.

Vulnerability is here regarded in a rather restricted sense: primarily as diminished autonomy due to either internal conditions (compromised consent) or one particular external circumstance (subordinate position).

The HIV/AIDS pandemic, first recognized as a new disease in 1981, necessitated CIOMS to develop new guidelines. At the same time, there was rapid growth of multinational research as well as increasing concerns about the involvement of vulnerable population groups, reinforced by fears for possible stigmatization of HIV patients. CIOMS therefore faced the challenge of providing clear guidance to protect vulnerable communities and individuals.

In 1991, the *International Guidelines for Ethical Review of Epidemiological Studies* were published. These guidelines articulate vulnerability in two ways. It is for the first time that vulnerability is specifically mentioned as a fundamental principle included in the general principle of respect for persons. It is argued that respect for persons incorporates two other fundamental principles. One principle is autonomy; the other is "protection of persons with impaired or diminished autonomy, which requires that those who are dependent or vulnerable be afforded security against harm or abuse."

The second articulation of vulnerability is as one item to consider in the ethical review procedure. Here the text enumerates several vulnerable populations, linking vulnerable and dependent groups in the same Art. 43: "Ethical review committees should be particularly vigilant in the case of proposals involving populations primarily of children, pregnant and nursing women, persons with mental illness or handicap, members of communities unfamiliar with medical concepts, and persons with restricted freedom to make truly independent choices, such as prisoners and medical students." (CIOMS 1991). Apparently the same populations as in 1982 are mentioned but also new groups are included.

Only a few years later, 1993, CIOMS published new Guidelines for Biomedical Research, superseding the 1982 Proposed Guidelines. The *International Ethical Guidelines for Biomedical Research Involving Human Subjects* are basically a statement of general ethical principles (the three principles of the Belmont Report) and 15 guidelines. It frequently mentions the notion of vulnerability. However, vulnerability is no longer identified as a fundamental principle. The notion is introduced in one of the first paragraphs of the document in connection to the principle of respect for persons with exactly the same wording as in 1991, however, as a fundamental ethical consideration, not a principle. Vulnerability is also associated with the principle of justice.

It is identified as a morally relevant distinction between persons that justifies a difference in distribution of burdens and benefits. The text provides a description of what vulnerability is: "a substantial incapacity to protect one's own interests owing to such impediments as lack of capability to give informed consent, lack of alternative means of obtaining medical care or other expensive necessities, or being a junior or subordinate member of a hierarchical group" (CIOMS 1993).

Vulnerability is also frequently mentioned in the commentaries to several Guidelines, specifying categories of vulnerable persons mentioned in previous guidelines, viz. children and persons unfamiliar with modern medical concepts (Guideline 1), persons with mental and behavioral disorders (Guideline 6), and prisoners (Guidelines 7). These are the same categories as mentioned in previous Guidelines, except that women are no longer associated with vulnerability. Guideline 10 now makes a distinction between 'traditional' classes of vulnerable individuals (mentioned in Guidelines 1, 6 and 7) and other vulnerable social groups. The first is characterized by limited capacity or freedom to consent, the second by subordinate social position. For this second group the wording is more or less the same as that in the 1982 Proposed Guidelines and similar examples are mentioned. However, beyond these paradigm cases, the text expands the number of vulnerable group considerably, based on the same rationale of dependency: "residents of nursing homes, people receiving welfare benefits or social assistance and other poor people and the unemployed, patients in emergency rooms, some ethnic and racial minority groups, homeless persons, nomads, refugees, and patients with incurable disease."

They all may have attributes resembling those of the other vulnerable classes, and therefore they are in need of special protection. In these cases, there is no internal limitation of autonomy but vulnerability is associated with external conditions that impact on autonomous decision-making (weakness due to social and existential conditions but also dependency).

The most recent revision of the CIOMS Guidelines in 2002 has the same structure as previous documents but more guidelines (21 instead of 15). It also puts more emphasis on human rights. In the Introduction, it is stated that human rights law in the application of ethical principles to research concerns primarily two principles: respect for autonomy and protection of dependent or vulnerable persons and populations (CIOMS 2002, p. 11). This seems to reiterate the position of the 1991 Guidelines that protection of vulnerability is a principle. But in the paragraph on General Ethical Principles, nothing has changed.

The text repeats that vulnerability is an ethical consideration when applying the principle of respect for persons. The text in relation to the principle of justice has also remained exactly the same. There is one new dimension. Vulnerability is not only relevant because it requires special protection, but it also demands special responsiveness: "Justice also requires that the research be responsive to the health conditions or needs of vulnerable subjects." (CIOMS 2002, p. 18). This positive approach is reflected in the commentary to Guideline 12 where it is stated that members of such groups have the same entitlements to access potential therapeutic benefits of research as non-vulnerable persons.

The major change in the 2002 Guidelines, however, is that vulnerability now is upgraded to a Guideline itself. Guideline 13 states: "Special justification is required for inviting vulnerable individuals to serve as research subjects and, if they are selected, the means of protecting their rights and welfare must be strictly applied" (CIOMS 2002, p. 64). There are also two changes that considerably enlarge the scope of the notion. One is that a broader definition is provided: "Vulnerable persons are those who are relatively (or absolutely) incapable of protecting their own interest. More formally, they may have insufficient power, intelligence, education, resources, strength, or other needed attributes to protect their own interests." (CIOMS 2002, ibidem).

Vulnerability no longer is a "substantial incapacity" but a relative quality that may result from a wider range of insufficiencies. The broader definition leads to a second change compared to previous guidelines: the number of vulnerable groups is growing. First, there is the group with limited capacity or freedom to consent (now called the 'conventional' group). Then, there are the groups that are vulnerable due to their social position of dependency. Third, there are simply 'other vulnerable groups' (the text of the commentary does not make a connection here with the social context). New groups concern elderly persons, displaced persons, individuals who are politically powerless, but also "countries or communities in which resources are limited to the extent that they are, or may be, vulnerable to exploitation…" (CIOMS 2002, p. 51). Finally, at least some women are back in the vulnerability framework: "…in some parts of the world women are vulnerable to neglect or harm in research…" (CIOMS 2002, p. 73).

3.2.3 The Declaration of Helsinki

The story about this authoritative statement of the World Medical Association (first adopted in 1964) can be much shorter since the Declaration of Helsinki did not mention the notion of vulnerability until its fifth revision in 2000. Vulnerability in this version is a consideration mentioned in the Introduction. The Declaration points out that some research populations are vulnerable and in need of special protection. Five groups are listed although it is not clear that they are all considered as vulnerable: "the economically and medically disadvantaged"; "those who cannot give or refuse consent for themselves"; "those who may be subject to giving consent under duress"; "those who will not benefit personally from the research"; and "those for whom research is combined with care" (WMA 2000).

The sixth (6th) revision of 2008 presents a reduced and changed statement regarding vulnerability in the Introduction. But at the same time, vulnerability is now mentioned under Principles (Art. 17): "Medical research involving a disadvantaged or vulnerable population or community is only justified if the research is responsive to the health needs and priorities of this population or community and if there is a reasonable likelihood that this population or community stands to benefit from the results of the research" (WMA 2008).

3.2.4 The Universal Declaration on Bioethics and Human Rights

In 2005 all member states of UNESCO unanimously adopted the *Universal Declaration on Bioethics and Human Rights*. This is the first statement of bioethical principles adopted by governments. It is part of international human rights law, although not legally binding. The Declaration states 15 ethical principles for global bioethics, including the three principles mentioned in the Belmont Report, although in slightly different formulation: autonomy and individual responsibility; benefit and harm; equality, justice and equity. In Article 8 the Declaration formulates the principle of respect for human vulnerability and personal integrity: "In applying and advancing scientific knowledge, medical practice and associated technologies, human vulnerability should be taken into account. Individuals and groups of special vulnerability should be protected and the personal integrity of such individuals respected" (UNESCO 2005).

Vulnerability is not explicitly defined but somewhat elaborated in various other articles. In the Preamble, the Declaration refers to the special needs of 'vulnerable populations'. In Article 24 (on international cooperation) it is emphasized that not only individuals may be rendered vulnerable, but also families, groups and communities. The Article furthermore refers to certain circumstances that may render individuals, families, groups and communities vulnerable, not listing specific vulnerable categories but explicitly mentioning conditions that may create vulnerability: disease, disability, plus other personal, societal and environmental conditions, and those with only the most limited resources.

It is not the first time that UNESCO documents refer to vulnerability. In the *Universal Declaration on the Human Genome and Human Rights* (adopted in 1997) vulnerability is mentioned in the context of international cooperation (Art. 17), emphasizing that states should respect and promote the practice of solidarity "towards individuals, families and population groups who are particularly vulnerable to or affected by disease or disability of a genetic character" (UNESCO 1997). Vulnerable groups are explicitly mentioned in Art. 24 as one of the parties concerned in the implementation of the Declaration; appropriate consultations with them should be organized. The 2005 Declaration obviously has built on these earlier formulations and has given them a much wider scope beyond the field of genetics. The *Interna-*

tional Declaration on Human Genetic Data, adopted by UNESCO member states in 2003 also refers to vulnerability but in a very specific context, referring to "special needs and vulnerabilities of developing countries" (UNESCO 2003).

It is important to note that in the Declaration adopted in 2005 respect for human vulnerability is promoted as one of the fundamental ethical principles for global bioethics. Although bioethics is not defined, the focus of the Declaration is wider than health research. As a principle vulnerability is relevant for healthcare, research and the application of technologies. Vulnerability is applied to individuals but also to families, groups, communities, populations. There are also groups of special vulnerability; no list of such groups is provided but specific conditions 'rendering' vulnerability are mentioned. Finally, explicit reference is made to human vulnerability. The focus therefore is on an anthropocentric interpretation of vulnerability. The point of reference is human life, not life in general. This is consistent with the scope of the Declaration which applies to human beings.

3.3 Controversial Dimensions of Vulnerability

The emergence of the notion of vulnerability in formal and authoritative bioethics documents highlights the following dimensions that have been the subject of intense scholarly debate since then:

a. The *status* of the notion of vulnerability has significantly evolved. In most of the documents it was first mentioned as a special consideration that needs to be taken into account in the application of general principles in research. Following the lead of the Belmont Report that introduced the notion of vulnerability in its discussion of the application of the three basic ethical principles for research, the CIOMS guidelines of 1982 as well as the Declaration of Helsinki in 2000 introduced vulnerability as a special consideration when applying general principles. However, the epistemological status of vulnerability was enhanced significantly in the CIOMS guidelines of 1991 that refers to vulnerability as a fundamental principle. However, the statement is ambiguous since it does articulate on the one hand that it concerns a principle, on the other that it is incorporated in the principle of respect for persons.

   The ambiguity continues to exist in subsequent CIOMS guidelines. In 1993, vulnerability no longer is a principle but rather a special application of principles. In 2002, vulnerability is mentioned at the same time as a principle in human rights law, an ethical consideration when applying the principle of respect for persons, as well as a guideline itself. Finally, the UNESCO Declaration clearly promotes vulnerability at the level of fundamental bioethical principles not only in research but also in health care. The latest revision of the Declaration of Helsinki in 2008 mentions vulnerability as one of its principles.

   The status of vulnerability has therefore been strengthened over the last decade. It is no longer a consideration within the application of ethical principles, or subsumed under another principle, but itself an ethical principle that needs to be balanced against other ethical principles such as respect for persons, beneficence, and justice. The reasons for this rising epistemological status are not clear. It is probably related to the globalization of bioethics, given the international status of the documents that have contributed to it. The CIOMS documents not only present a global perspective but also articulate the human rights perspective. Another possible reason is the long-time dominance of the discourse of individual autonomy as characteristic for the Western context in which bioethics has emerged.

As long as vulnerability has been defined as limitation of autonomy, it can only be considered as a special consideration or special application of the fundamental principle of respect for autonomy. Vulnerability could only be conceptualized as a separate principle if its contents were emancipating from the autonomy discourse.

b. The second issue is the *content* of the notion of vulnerability. In many cases vulnerability is not defined. Individuals and groups are labeled as vulnerable but it is not always clear what the criteria are for doing so. The Belmont Report relates vulnerability to dependency and compromised capacity to consent. In the background papers to the Report Robert Levine pays attention to the notion of vulnerability.

He identifies vulnerable persons as a subcategory of persons not capable of protecting their own rights and welfare. Vulnerability therefore is a special class of limited or impaired autonomy. But it is not the same as lack of comprehension and dependent status that may also compromise autonomy and autonomous decision-making. Levine seems to imply that vulnerability emerges because of the special condition of the individual or group of individuals. His examples, viz. the sick and the impoverished, do not refer to dependency or lack of comprehension but to the circumstances that might impede really free decision-making.

The ambiguity about the criteria and thus the content of vulnerability is reflected in the 1982 CIOMS guidelines. Two criteria are used here: compromised consent and dependency (specifically persons who are in a subordinate position in a hierarchy). The 1993 CIOMS guidelines for the first time present a description of the notion of vulnerability: “a substantial incapacity to protect one’s own interests.” They also distinguish three subcategories. The first (lack of capacity to give informed consent) identifies what has become known as the traditional class of vulnerable persons. The second (lack of alternative means to obtain medical care) refers in fact to the social position that renders persons vulnerable, while the third (subordination in a hierarchical group) emphasized the criterion of dependency.

This conceptual approach is reminiscent of the earlier analysis of Levine but restructures it. Vulnerability is now the name of the broad category and it includes the subcategories of lack of comprehension and dependency. At the same time, the notion is now moving beyond the perspective of limited autonomy. The above description of vulnerability is literally repeated in the 2002 CIOMS guidelines. But in another section it is slightly reformulated, not as "substantial incapacity" but as "relatively (or absolutely) incapable."

It is also expanded with references to possible explanations of vulnerability: insufficient power, intelligence, education, resources, and strength may make people unable to protect their own interests. This formulation again, is very similar to the one used by Levine (1981, p. 54) in the first edition of his textbook on research ethics. Levine has been intimately connected with the drafting of the CIOMS guidelines, as co-chair respectively chair of the Steering Committee in 1993 and 2002, and it seems that his ideas on vulnerability have been broadly accepted. The implication of the reformulation is that the scope of vulnerability is enlarged; it is a relative quality resulting from a range of insufficiencies, both internal and external, and not limited to impaired autonomy.

c. The third dimension is the *scope* of vulnerability. The notion is used to qualify individuals, but also families, groups, communities, populations, and even countries. It is often assumed that vulnerability is transferable from one category to another. What has started as examples of vulnerable groups has resulted in growing lists of vulnerable populations. Gradually, consensus has emerged on some criteria and categories, for example, limited capacity or freedom to consent (and thus, children, persons with mental or behavioral disorder, and prisoners) and social position of dependency (thus, subordinate members in a hierarchy, the institutionalized, and the poor).

However, the list of groups included in the second category has steadily grown. The identification of a third category ('other vulnerable groups') has opened the door to the labeling of even more groups as vulnerable, especially the elderly and women. This broadening scope has made the notion of vulnerability not only increasingly difficult to apply in practical settings but has also undermined its relevancy. If almost every person is a member of one or more vulnerable groups what will be the moral significance of identifying lists of vulnerable categories?

The 2002 CIOMS guidelines present the most encompassing classification of vulnerable groups. Since then, a different approach can be discerned. The Declaration of Helsinki in 2008 only uses vulnerability for populations and communities. In distinction to the 2000 fifth revision which lists five categories of vulnerable populations, the sixth revised text only briefly mentions, by way of example, two groups. The UNESCO Declaration does not list vulnerable categories at all but merely mentions conditions that may produce vulnerability. It is recognized that vulnerability may be manifested at various levels (individual, family, group, etc) and that 'groups of special vulnerability' can be distinguished, but a simple categorization is impossible. Vulnerability can be the result of internal conditions (impaired autonomy) or external conditions (dependency and weakness) but more often of interactions between these conditions, posing

threats to some individuals or groups of individuals rather than whole categories of persons.
The identification of vulnerability should be subtle and nuanced. Threats can best be identified at the level of the individual (as part of a group or community) since the Declaration associates vulnerability with the personal integrity of individuals. This more subtle approach of identifying conditions of vulnerability rather than classes of vulnerable individuals is also accommodating the critique that classification of vulnerability may result in discrimination and stigmatization.

d. Finally, it is debatable what exactly the moral implications of vulnerability are. This is the dimension of the practical *consequences* of the notion of vulnerability. At least three different consequences are pointed out in the documents. First, the need of extra justification for involving vulnerable persons and groups. This implication is emphasized in the Belmont Report. The assumption is that vulnerable populations are easy to recruit for research; the appropriateness of involving them should be specifically demonstrated and justified. The 1991 CIOMS guidelines take a similar approach. They require that ethical review committees are particularly vigilant when vulnerable populations are involved.

   A second consequence is that vulnerability demands special protection. Because vulnerable persons cannot protect their own interests, others will be responsible for protecting their interests. This is the implication articulated in the 1993 CIOMS Guidelines and the 2000 Declaration of Helsinki.

   The third consequence formulates a more positive approach. The 2002 CIOMS guidelines point out that vulnerability demands that research should be responsive to the needs of vulnerable persons. This responsiveness and probability of specific benefits of research with vulnerable populations is even more stringently articulated in the latest revision of the Declaration of Helsinki. The UNESCO Declaration also relates vulnerability to a positive duty. States should promote solidarity and international cooperation in order to answer to the special needs of vulnerable populations.

   Although the three consequences are different, they cannot simply be characterized as pointing towards negative or positive duties. The underlying assumption is that vulnerable populations can be involved in research but that extra caution is necessary. Researchers are not asked to refrain from including vulnerable subjects but they should provide special justification and special protection. To a certain extent restraint is required (negative duty) but at the same time, if they are included, additional activities are requested (positive duty). The most recent documents put more emphasis on the positive demands.

   Vulnerability is not merely a condition to take into account but it is an incentive to help people cope with their vulnerable condition. This emphasis is related to the application of the notion to health care in general and not just to the area of medical research. Respect for human vulnerability formulates a normative prescription to care for those threatened by biological and social circumstances, as well as by the power of medicine itself. It stipulates that bioethics is more than respecting individual choices and personal autonomy: it aims at safeguarding care for other persons.

3.4 Conclusion

Vulnerability is a complex notion in need of further analysis and clarification. Being introduced in the bioethical discourse via the Belmont Report in 1979 it has played an increasingly important role in international policy documents, culminating in 2005 in its formulation as a separate bioethics principle. The principle has been criticized as being too vague, too broad, too narrow, and futile since it does not provide clear moral guidance.

Nevertheless, in the context of global bioethics, the language of vulnerability is indispensable (Solbakk 2011).

On the one hand, vulnerability is a general characteristic of the human condition. It expresses the fragility and finitude of human existence, and is therefore a feature shared by all human beings. On the other hand, vulnerability refers to the fact that some people are more vulnerable than others, due to natural or social conditions such as disease or poverty. It expresses the normative requirement that these vulnerable fellow human beings need special care.

More is needed than non-interference; they should receive assistance that will enable them to realize their potential as human beings.

References

CIOMS. 1982. Proposed international guidelines for biomedical research involving human subjects. In *Bioethics. issues and perspectives,* ed. S. Scholle Connor and H. L. Fuenzalida-Puelma, 220–226. Washington, DC: Pan American Health Organization.

CIOMS. 1991. International guidelines for ethical review of epidemiological studies. Sahih. http://www.cioms.ch/publications/guidelines/1991_texts_of_guidelines.htm. Accessed 1 April 2012.

CIOMS. 1993. International ethical guidelines for biomedical research involving human subjects. Sahih. http://www.codex.uu.se/texts/international.html. Accessed 1 April 2012.

CIOMS. 2002. International ethical guidelines for biomedical research involving human subjects. Sahih. http://www.cioms.ch/publications/guidelines/guidelines_nov_2002_blurb.htm. Accessed 1 April 2012

Levine, Robert J. 1981. *Ethics and regulation of clinical research.* Baltimore: Urban & Schwarzenberg.

Solbakk, Jan Helge. 2011. Vulnerability: A futile or useful principle in healthcare ethics? In *The Sage handbook of health care ethics: Core and emerging issues,* ed. Ruth Chadwick, Henk ten Have, and Eric M. Meslin, 228–238. London: Sage.

The Belmont Report. 1979. Ethical principles and guidelines for the protection of human subjects of research, report of the National Commission for the Protection of Human Subjects of Biomedical and Behavioral Research. *Federal Register* 44 (76): 23191–23197.

The Belmont Report. Appendix Volume I & II. 1978. National Commission for the Protection of Human Subjects of Biomedical and Behavioral Research. DHEW publication no. (OS) 78-0013.

UNESCO. 1997. The universal declaration on the human genome and human rights. Sahih. http://unesdoc.unesco.org/images/0012/001229/122990eo.pdf. Accessed 1 April 2012.

UNESCO. 2003. International declaration on human genetic data. Sahih. http://portal.unesco.org/en/ev.php-URL_ID=17720&URL_DO=DO_TOPIC&URL_SECTION=201.html. Accessed 1 April 2012.

UNESCO. 2005. Universal declaration on bioethics and human right. Sahih. http://unesdoc.unesco.org/images/0014/001461/146180e.pdf. Accessed 1 April 2012

WMA (World Medical Association). 2000. Declaration of Helsinki, 5th revision. Sahih. http://ohsr.od.nih.gov/guidelines/helsinki.html. Accessed 1 April 2012.

WMA (World Medical Association). 2008. Declaration of Helsinki, 6th revision. Sahih. http://www.wma.net/en/30publications/10policies/b3/index.html.pdf?print-media-type&footer-right=[page]/[toPage]. Accessed 1 April 2012.

Chapter 4
Vulnerability: Considerations on the Appropriate Use of the Term in Bioethics

Martha Tarasco Michel

4.1 Introduction

This paper poses a question on the use of the term "vulnerability." Is this polysemous word often abused today[1]? Is it an analogical term? Or does it have only one meaning?

The expression "vulnerable groups" is mentioned and used in many fields, often in reference to human rights. It is used particularly in situations of economic marginality, in terms of sexual preference, or most commonly, when referring to women. The term is also used when seeking to protect the terminally ill, subjects of clinical trials, or human embryos. It is, therefore, worth examining the use of this term in different areas. In this paper the author intends to reveal the existence of certain abuses in the use of this term.

Several theses will be argued for throughout the various sections. The main assertion, however, will be controversial. It has to do with the fact that certain uses of the term "vulnerability," from a strictly bioethical point of view, are confusing linguistic conflations. From the author's point of view, it would perhaps be more accurate to say that there is a certain progression from a legitimate use of the term to an inconsistent one.

[1] Polysemy (from Greek, poli-, many, and σῆμα, meanings) is present when the same word has many meanings.

M. Tarasco Michel (✉)
Faculty of Bioethics, Anáhuac University, Mexico City, Mexico
e-mail: mtarasco@anahuac.mx

J. Tham et al. (eds.), *Religious Perspectives on Human Vulnerability in Bioethics*, Advancing Global Bioethics 2, DOI 10.1007/978-94-017-8736-9_4,

4.2 Some Uses of "vulnerability"

There are several uses of the term in ordinary language. Although they are always in reference to a risk, a threat, or danger, semantic contexts and meanings differ. There is always a reference, implicit or explicit, to something or someone else. It can even be a mere hypothetical assumption:

Subjective but realistic perception	Hypothetical perception
"I feel vulnerable before you."	"Vulnerability of the operative system."
(*since you make me feel weak and/ or you expose my weaknesses to others*)	*For example, a government that cannot be sustained any longer*
"The body is vulnerable."	"The economy is vulnerable."
There are physiological limits and illnesses that affect the body	*For example, if Greece cannot remain in the Euro monetary zone, other economies of the European Union will be damaged*
"Women are more vulnerable than men."	"The defense system is vulnerable."
In terms of muscular strength, women are generally weaker than men even though they may—be stronger emotionally	*When the Twin Towers and the Pentagon were attacked, the American defense system was shown to be weaker than previously believed*

We also see that "vulnerable" can be applied to groups or to individuals. Moreover, it is evident that the term is always applied to an element, a function, or a structure which may not work when they are compared to one another.

When analyzing some examples of theoretical descriptions and definitions (as in "Victimization and vulnerability due to lack of legislation and other measures relating to domestic violence," Pérez 2005). Maria Montserrat Pérez Contreras says:

> … although it is true that domestic *violence can be brought against any member of the family*, it is women, along with children, who *most often suffer* multiple violations to their rights as persons in the household. *That is*, women are the main victims and belong to the *most vulnerable group* regarding the phenomenon of domestic violence (ibidem).

According to this description, vulnerability is in fact present in all of us, and the issue of using the adjective or of including someone in the "vulnerable" group is a quantitative and cultural issue. In other words, we see a confirmation of the proposed characteristics in the aforementioned examples in the table above, as well as in those italicized in the citations. For example, vulnerability occurs in relationships—in this case, in the family. It is always in relation to another and is a matter of degree. For, as the author recognizes, in a certain way we are all vulnerable, or at least at risk of becoming so.

The same author confirms this approach elsewhere. Citing the Mexican National Human Rights Commission, she points out:

> Vulnerability stems from the association of internal and external factors which, when combined, reduce or abrogate the ability of a person, group, or community to face a given situation that harms them and, in addition, to recover from it. (ibidem.)[2]

[2] Note that the definition adds an additional component: the notion of damage or harm. It is clear that the expression does not have the sense of "I am vulnerable to winning the lottery," or "She is vulnerable to being healthy," etc.

We get the impression that there is a latent risk of turning apparent vulnerability (which given the use of the language would suggest a social category) into a supposedly new metaphysical category. If that position is accepted, one could then arrive at false generalizations of the concept, with statements such as, "All children are vulnerable," or "Children are by nature more vulnerable than adults." We should ask: *what* are they vulnerable to?

Of course, this paper does not ignore the body's vulnerability or the fact that human existence itself is always imperfect both in time and in its functions. And besides being imperfect, there is always a chance of dying or losing vital functions. This concept is really part of the human condition (Urdanoz 1991; Beuchot 1992; Nédoncelle 1948; Lobato 1994; Millán 1985; Llano 1995; Aquinas 2010, I.art.3).[3]

It is an objective fact that children belonging to certain groups or having certain conditions are more vulnerable (Kelly et al. 2003; Bates et al. May 2004; Perrin et al. March 1989). But the same can be said of adults as much as children. During a measles epidemic, an adult is more vulnerable than a child to suffer death or recurrence of the disease (Salamá et al 2001).

An adult is more vulnerable than a child to permanent damage from a bone fracture (Mahncke et al. 2006).

An adult has less brain plasticity than a child, making him more vulnerable to permanent brain damage after a head injury. Seen from this perspective, the category of "vulnerability" adds nothing new to the traditional category of "harm" and "benefit" (Beauchamp and Childress 1994; Ross 1994; Rawls 1971; Sgreccia 2009), terms traditionally used in medicine and bioethics.

So why should we not use the term in bioethics? In an article entitled "Bioethics of Protection, Vulnerability, and Threat from a Secular Perspective", the term "vulnerability" appears near the end of the article:

> The examples analyzed above [the author reviews cases of human embryos] refer exclusively to living or potential human beings, or to 'parts' of them. The analysis of protective secular bioethics is appropriately applied, in principle, to all of them, as all of them are in some way beings or entities *considered and/or threatened under the term vulnerability* (schramm March 2003, emphasis added).

It has been emphasized that the use of the term "vulnerable" is an approach or a category used in a relationship which relatively indicates when one is in a vulnerable situation or not. In other words, there is no intrinsic vulnerability; there are only conditions which, as a result of personal actions or those of somebody else, put this person at greater risk in comparison to other persons or circumstances. Hence, vulnerability as mentioned is always a concept of comparison, either among the past, present or future states of vulnerability, or among groups, or among individuals or circumstances.

Combining all these elements, we can propose another definition of vulnerability that seems to summarize what has been said so far:

[3] Here we mean true human liberation. Man always perceives himself as a contingent being who yearns for the Absolute. Good and Being refer to the same reality. From here is the link between being as perfection and as action.

> The term vulnerability, although it includes poverty and exclusion, is broader than these because it includes a larger number of risks. Vulnerability is the propensity to injury from a certain potentially destructive force or energy. It is the inability to absorb the effects of a given change and to adapt to these modifications. (Foschiatti 2007)

The category of vulnerability, then, belongs to the sphere of medicine, since, under similar conditions, one organism may be more susceptible than another to acquire a certain type of disease (Singer 2002; Prins 2004; Feldbaum et al. 2006; Frich et al. 2006). Nevertheless, in any medical condition, the term vulnerability is merely a synonym for risk. When we single out a group or an individual, we say they are "vulnerable." When I was researching the medical literature on this concept, "vulnerability" is always used in place of "risk" even when it is explicitly a subjective perception of risk. We should therefore wonder if the concept is useful in the field of bioethics.

"Vulnerability" is also used as synonym of "fragility." Fragility and vulnerability imply or suggest the idea of incompleteness, of something left unfinished. But if this semantic interpretation is accepted, we would have to conclude that vulnerability is permanent. This would be so because a person is always incomplete, always fragile in all expressions and areas of his corporeality. This is so because a person never reaches his full potential (Kierkegard 1944; Sánchez-Magallón 2003; Burgos 2000; Ebner 1995).

It also seems that when the term is used for an association, the group thus formed becomes fragile, vulnerable or susceptible to failure before a real or imaginary threat. For example, when one says that women and children are vulnerable, it implicitly means that they are fragile. Similarly when the term "vulnerable" is used in the field of healthcare, it could equally be substituted with "being at risk." We could often find other expressions that are equivalent to the term in question (Barwise and Perry 1983).

It is suggested that when the term "fragility" is used, it refers to the 'order of being', that is to say "essentially." Vulnerability, on the other hand, refers to "the order of situational being", that is to say temporally or circumstantially. Their meanings, however, are sometimes similar or even interchangeable. At times they are also equivalent in some way, since both share the idea that a certain limitation is perceived for a lack of "something"; and this without ignoring what remains of the conceptual content of either term.

The human person is, therefore, always fragile and always vulnerable. Fragility persists in every circumstance. Any man, woman or child with no food, oxygen, etc., will die. They can be injured. They are fragile. Vulnerability, however, is circumstantial. Persons become vulnerable because their fragility is pushed to the limit. It seems that this is why these two expressions are not equivalent. The expression "this man is fragile" is not entirely equivalent to "this man is vulnerable."[4] Strictly speaking, we are always fragile. Fragility refers to our natural state and its inherent limitations. But the second term refers to an external relation. We have a clear example in the use of the verb "to be" as in "to be someone" vs. "to be" as in "to be sick" (in Spanish these are two different verbs, "ser" and "estar"). For example,

[4] Fragile is defined in many dictionaries as a synonym of weak, inconsistent, slack, feeble, and delicate. Its antonyms are consistent, strong, and solid.

compare the expression "the patient is in a delicate condition" to the expression "the patient is delicate."[5] Being in a delicate condition is circumstantial—the patient has become vulnerable—but he will always be fragile.

In a radical sense, we live in a state of permanent vulnerability. The entire human race, limited by its own corporeality, lives under the threat of harm from natural and social phenomena See Beth 1961; Prini 1991; Entralgo 1991; Arregui 2006; Aquinas 2010. People are not only fragile; they are, in a certain sense, always vulnerable. When using this term in everyday language, though, we refer to the nearest, most probable and most measurable threats. Someone on a soccer field in a thunderstorm is, for instance, already fragile, but his vulnerability is greater because of the danger of being struck by lightning.

Something similar happens in bioethics. In the field of medicine we humans are always fragile. Health is a state of equilibrium in constant risk of failing. Vulnerability is more of a social notion. It is perhaps useful when speaking in terms of population policies and other similar areas. In the clinical area of medicine, however, the term is broadened to include the social dimension. In terms of clinical practice, one harms or does not harm, one heals the one who suffers, etc., but one does not strictly combat vulnerability. Moreover, it appears that the use of this term (so popular in recent times, but one that did not originate in medical practice) leads to certain actions which, from an ethical point of view, do not correspond to the true purpose of medicine. The following paragraphs will develop this idea.

4.3 A Key to Interpretation: Vulnerability and Corporeality

Corporeality signifies "being one with the body which means it cannot be conceived as just another resource to be used like the way natural resources are used."[6] While this term gives the sense of what the human body has, it is important to emphasize that even when we seek the unlimited, the body is in itself limited.

As already stated, this is true both anatomically and physiologically—our arms are not designed to walk permanently, we cannot survive for more than a few minutes without breathing, every organ or system cannot perform optimally at all times, and youthfulness cannot be prolonged forever. This single fact should convince us of the fragile condition of all humanity. As the body cannot survive more than three days without being hydrated, everyone is vulnerable to the lack of water. But with intelligence, we can find water even in the desert. So while our body is vulnerable, with the use of our intelligence, which subsists in our corporeality, we can overcome these vulnerabilities.

Naturally, since we are multi-tasking organisms, we can use these concepts of fragility and vulnerability without difficulty because in a certain sense they have

[5] The Spanish verb *ser* indicates to be in a permanent state, while *estar* means a changeable state. Still there can be many exceptions.

[6] This idea is, however, contrary to dualistic anthropology. See Singer 1994; Grisez 1977.

very similar meanings. Total vulnerability does not exist. The term is always used as part of a more or less complex element or system, in which a part or an element of the system is exposed. When this is used as a concept for an "absolute state," it acquires an ideological sense—that is, it is not gained from experience. Normally, when this term is used properly, that is, not ideologically, it becomes intelligible.

The term vulnerability can be used for rhetorical purposes which can serve as a "wake-up call" or a warning signal when someone is facing failure and destruction or irreparable damage is imminent. This applies whether the failure is apparent or real, and whether it affects a person, an institution, or a thing. This failure, though, is due to some inherent weakness present. This seems to be the sense intended in various statements and declarations that seek to defend "vulnerable groups"[7]—groups said to be in complete disadvantage or completely abandoned, etc.—with no proof of the objective truth of such a statement.

Legislation aimed at protecting vulnerable groups emphasizes their specific weaknesses. It focuses especially on weaknesses of cultural origin that seek an adjustment of the political will so as to avoid a structural imbalance. Take, for example, a public health policy aimed at a "vulnerable" group of poor people by providing them free vaccinations. Such a policy does not imply that those who can pay for the vaccine in private medical practices will have better immunity. Both groups are equally vulnerable to become infected, but their vulnerability is different in that their likelihood of exposure is different due to their economic conditions. Hence, labeling a specific group with the term "vulnerable" may not show the complete picture. A rich or a poor person is equally vulnerable to an infection. If a public policy proposes to give free vaccination to a vulnerable group, at least an adjective should be added to specify what kind of vulnerability: economic status or health status.

4.4 Vulnerability and Desire: A False Source of Vulnerability

Taking the previous section's assertions into account, some elements outside human fragility are nevertheless presented as vulnerability at times. One source of these paradoxes is the notion of autonomy, understood as "stripping down" or even suppressing the principle of corporeality.

Personal autonomy is seen as the desire to choose in matters related to one's own corporeality, as if the body, itself considered just one more object of nature, were a personal choice intended to transform nature. An example illustrating this would see nature or body as "transformable" by desire and the possibility of self-realization.

[7] "The question of corporeality is also decisive in ethical theories in general… Everything changes in ethics, insofar as the person is—based on Kant—considered a free and responsible moral agent… With this perspective, having more freedom, dominating nature, is not only good but is even the ideal of excellence. Derechos de las Mujeres, Normativa, Interpretaciones y Jurisprudencia Internacional. Secretaria de Relaciones Exteriores. Oficina en México del Alto Comisionado de las naciones Unidas para los Derechos Humanos (2006). UNIFEM, p. 220.

We then have situations of apparent vulnerability. Mere desire would make us "vulnerable." One could think of "my autonomy" or "respect for my choice." As long as it is "my choice," it could lead to false hopes or an ambiguous idea of fulfilling a legitimate human aspiration. That is, if some choices are not allowed, then the person is then in a state of "vulnerability." Suppose someone wants to have children and tries every available means to achieve this goal, but is still prevented from achieving it. This could be interpreted as "vulnerability" in the area of reproduction. For example, "some people have a kind of control that I do not have, and they are hindering or violating my freedom…" Nevertheless, what we desire should perfect our being. When we recognize that not everything we desire is good, that recognition does not make the will vulnerable or subject it to harm; rather, it helps to perfect one's own corporeality—not in the sense of perfect functionality, which is never achieved, but in the sense of the ultimate purpose of corporeality.

In simpler terms, if we call a subject "vulnerable" just because he or she cannot do or attain anything he or she wants, it becomes a meaningless expression.

Even if not everyone agrees, the fact that vulnerability is relative does not mean it is arbitrary. There are minimum objective criteria useful for identifying it. These criteria include the possibility of objective harm: real and verifiable, not just perceived or hypothetical. "A child is vulnerable simply because he cannot make certain decisions that an adult can make." This is not vulnerability. The vulnerability that can affect a child, a woman, or anyone else is the possibility of not being able to avoid danger or recover from harm.

This is defined not only by a personal choice, but by the decisions of others and what the first author cited in this paper calls "external factors." The "internal factors" are certainly not mysterious mental states, but actions in which the subject could suffer harm.

4.5 Conclusion

1. Vulnerability implies a temporary, situational category while fragility places more emphasis on limited human nature.
2. From a medical point of view, there are vulnerable groups if the term is employed to mean susceptibility or propensity to disease. Every organism is fragile, which is different from clinical vulnerability.
3. Men and women in any stage or condition of life are always vulnerable in relation to different circumstances or approaches, but it is not an ontological state. Rather, it is a state of existence (in terms of time, limitations, fragility, characteristics of culture, age, and sex) which is not constant but is present in specific moments.
4. Autonomy and vulnerability are not the same, and one is not the cause of the other. On the contrary, autonomy detached from the objective reality of the subject may put him in a very vulnerable situation based on the outcome of him own desires. In this sense, these two terms might even be seen as antagonistic.

5. In certain situations, age or gender groups may be more vulnerable than other groups. But the term vulnerability in this sense is used in comparison with other less vulnerable groups.
6. The term vulnerability is never used in the positive sense: no one, for example, considers himself "vulnerable" to receiving a prize. In this sense, it is different from the concept of susceptibility.

References

Aquinas, T. 2010. Summa Theologiae I-II Prologue

Arregui, J. V. 2006. Corporalidad. In *Diccionario de Bioética*, ed. Vázquez. Monte Carmelo: Burgos.

Bates, I., C. Fenton, and J. Gruber 2004. Vulnerability to malaria, tuberculosis, and HIV/AIDS infection and disease. Part 1: Determinants operating at individual and household level. *The Lancet Infectious Diseases* 4 (5).

Barwise, J. K., and J. Perry. 1983. *Situations and attitudes*. Cambridge: MIT Press.

Beauchamp, T. L., and J. F. Childress. 1994. *Principles of biomedical ethics*. Oxford: Oxford University Press.

Beth, E. W. 1961. Semantics of physical theories. In *The concept and the role of the model in mathematics and natural and social sciences*, ed. Freudentahl. The Netherlands: Reidel.

Beuchot, M. 1992. *Introducción a la Filosofía de Santo Tomás de Aquino*. Mexico: UNAM.

Burgos, J. M. 2000. *El Personalismo*. Madrid: Biblioteca palabra.

Ebner, F. 1995. *La Palabra y las Realidades Espirituales*. Madrid: Carrapos.

Entralgo, L. 1991. *Cuerpo y Alma: Estructura Dinámica del Cuerpo Humano*. Madrid: Espasa-Calpe.

Feldbaum, H., K. Lee, and P. Patee, et al. 2006. The national security implications of VIH/AIDS. *PLoS Medicine* 3 (6).

Frich, J., L. Ose, and K. Malterud. 2006. Perceived vulnerability to hearth disease in patients with familiar hypercholesterolemia: A quantitative study. *Annals of Family Medicine* 4.

Foschiati, A. M. 2007. Vulnerabilidad, Pobreza y Exclusión. Problemas de Gran Impacto en el Nordeste Argentino. *Anales de Geografía* 27 (2).

Grisez, G. 1977. Dualism and the morality. In *Atti del Congresso Internazionale su Settimo Centennario di San Tommaso d'Aquino, 5,* ed. L´Agire Morale. Naples: Edizione Domenicane.

Kelly, A., K. Gil, and L. Schamberg. 2003. Brief report: Parental perceptions of child vulnerability in children with chronic illness. *Pediatric Psychology* 28 (3).

Kierkegard, S. 1944. *The concept of anxiety*. Princeton: Princeton University Press.

Llano, C. 1995. *Las Formas Actuales de la Libertad*. México: Trillas.

Lobato, O. P. A. 1994. *El Ser Personal*. Rome: Angelicum Preambulus.

Mahncke, H. W., A. Bronstone, and M. M. Merzenich. 2006. Brain plasticity and functional losses in the aged: Scientific bases for a novel intervention. *Progress in Brain Research* 157.

Millán, A. 1985. *Fundamentos de Filosofía*. Madrid: Rialph.

Nédoncelle, M. 1948. Prosopon et Persona dans l´antiqué Classique. Essai de Bilan Lingüistique. *Rev des Sciences* 22.

Pérez, M. 2005. Victimización y Vulnerabilidad por Ausencia de Legislación y Otras Medidas en Materia Familiar. *Boletín Mexicano de Derecho Comparado* 112.

Perrin, E., P. West, and B. Culley March 1989. Is my child normal yet? Correlates of vulnerability. *Pediatrics* 8 (3).

Prini, P. 1991. *Il Corpo che Siamo*. Turin: SEI.

Prins, G. 2004. AIDS and global security. *International Affairs* 80 (5).

Rawls, J. A. 1971. *Theory of justice*. Cambridge: Harvard University Press.
Ross, W. D. 1994. *The right and the good*. Oxford: Clarendon Press.
Salamá, P., A. Fitsum, and L. Talley, et al. 2001. Malnutrition, measles, mortality, and the humanitarian response during a famine in Ethiopia. *JAMA* 286 (5).
Sánchez-Magallón, S. 2003. *El Personalismo Ético de Dietrich Von Hildebrand*. Madrid: Rialph.
Schramm, R. 2003. Bioética de la Protección, Vulnerabilidad y Amenaza Desde una Perspectiva Laica. *Diálogo Político* 20 (1).
Sgreccia, E. 2009. *Manual de Bioética*. Madrid: B.A.C.
Singer, P. 2002. AIDS and international security. *Survival* 44 (1).
Singer, P. 1994. *Rethinking life and death: The collapse of our traditional ethics*. New York: St Martin's Griffin.
Urdanoz, T. 1991. *Historia de la Filosofía IV*. Madrid: B.A.C.

Part II
Religious Perspectives of Vulnerability from Philosophical, Ethical and Legal Points of View

Chapter 5
Vulnerability, Compassion, and Ethical Responsibility: A Buddhist Perspective on the Phenomenology of Illness and Health

Ellen Y. Zhang

5.1 Introduction

Perhaps no religion like Buddhism speaks explicitly of the human experience of bodily/mentally vulnerability. According to the Buddhist tradition, the story of the young prince Siddhartha and his decision of embarking on the journey of enlightenment is closely related to his personal experience of seeing the "four passing signs", three of which are about the experience of human physical vulnerability: aging, sickness, and death.

First Passing Sign: The young prince, who had been protected by the family from the sights and sounds of human suffering for 29 years of his life, was out in his chariot one day when he saw an older man, who "bent like a roof-beam, broken, leaning on a stick, tottering, sick, his youth all vanished" (*Anguttara Nikaya*).[1]

Deeply disturbed by what he saw, realizing that all humans, including himself, cannot escape from the miserable condition of losing youth and getting old, the young prince wondered, is anyone able to acknowledge that happiness and comfort such as he experienced in the palace is transient and temporary? The suffering of aging lies in wait for every human being.

Second Passing Sign: Some days later, the young prince was being driven around the royal park when he encountered "a sick man, suffering, very ill, fallen in his own urine and excrement, and some people were picking him up, and others putting him to bed." (*Anguttara Nikaya*).

Upon seeing the sick man shivering with fever and being told that poor man may not recover from the illness, the young prince went back to the palace. He thought to himself: "I, too, am subject to illness and cannot escape it. If I, who am subject

[1] The account of the story of the four passing signs is found in *Digha Nikaya* 1995, 14.2, pp. 207–210.

E. Y. Zhang (✉)
Centre for Applied Ethics, Hong Kong Baptist University,
Hong Kong, People's Republic of China
e-mail: eyzhang@hkbu.edu.hk

J. Tham et al. (eds.), *Religious Perspectives on Human Vulnerability in Bioethics*, Advancing Global Bioethics 2, DOI 10.1007/978-94-017-8736-9_5,

to illness without escape from it, should see another who is ill, and should be oppressed, beset and sickened, it would not be well with me." (*Anguttara Nikaya*).

Third Passing Sign: One day when the young prince was being driven around the royal park again, he saw a corpse being carried to the burning ground. Distressed by what he saw, the young prince asked: "But am I too then subject to dying, not exempt from dying? … I, too, am subject to death and cannot escape it. If I, who am subject to death without escape from it, should see another who is dead, and should be oppressed, beset and sickened, it would not be well with me." (*Anguttara Nikaya*).

After experiencing the above three signs, the young prince encountered a wandering religious beggar on his way to the royal park, which is called the "Fourth Passing Sign. The young prince was surprised by the beggar's countenance which was filled with peace and joy, and so he resolved to become also a wandering beggar … And we all know the rest of the story.

Of course the story of the "four passing signs" should be understood allegorically rather than as a faithful account of a historical event. Nevertheless, it tells us that Buddhism begins with specific existential questions engendered by human experience; that is, old age, illness, and death, the important aspects of human sorrow and suffering. As a matter of fact, Buddhism not only realizes the physical pains of human beings in terms of old age, illness, and death, it also talks about the mental anguish of human beings in terms of thirst, craving, and desire.

The word "suffering" is a translation of the word *dukkha* (Pali) or *duhkha* (Sanskrit), which literally means dis-ease or unsatisfactoriness. There is a well-known Buddhist claim, "All this is *dukkha*." Suffering is, then, shown as a kind of dis-ease caused by human limitation and finitude. Yet suffering is more complicated than a subjective, psychological description or a state of mind in terms of an intentionalist view in that the phenomenal character of any experience is entirely constituted by its representational content; instead, it has a wide range of meanings derived from reality itself, although Buddhism does not seem to focus on reality as it is apart from human experience.

The whole business of Buddhist enlightenment as such is about self-reflection on the experiential dimension of human life in reality as well as upon a possible way of releasing human beings from the experience of sufferings caused by "disease." The central tenet of Buddhist teaching is called the "four noble truths" which state there is:

1. The truth of what suffering is;
2. The truth of the conditioned cause of suffering;
3. The truth that suffering can be eliminated by eliminating the conditioned cause;
4. The truth of the means to eliminating suffering by following the Middle Way, constituted by the Noble Eightfold Path.

With regard to the truth of suffering, the Buddhist teaching further explains it as follows:

> The noble truth of suffering is this: birth is suffering; aging is suffering; sickness is suffering; death is suffering; sorrow and lamentation, pain, grief and despair are suffering; association with the unpleasant is suffering; disassociation from the pleasant is suffering; not to

> get what one wants is suffering—in brief, the five aggregates (*skandhas*) of attachment are suffering (*Anguttara Nikaya*).

Thus, suffering is the fundamental condition of the life of all sentient beings, regardless of a person's social status. Since human life is pervaded by suffering, both visible and invisible, what is needed is a remedy to cure the fundamental human illness of suffering. Buddhism, therefore, focuses on meditative insight into dependent origination, the ever-changing and interrelated nature of existence. This insight, which is also called wisdom (*prajña*) in the Buddhist teaching, provides a means for developing an awareness of all kinds of dispositional influxes that produce the illusion of a fixed self characterized by a mind of craving, greed and hatred. Vulnerability, according to Buddhism therefore, is not merely a physic-psychological description, pointing to corporeal limits or emotional frangibility, but a total and lived experience of human existence conditioned by the "four marks" of arising, abiding, altering, and perishing.[2] At the same time however, the world of suffering (*samsāra*) constitutes the experiential starting point of the soteriological possibility.

5.2 The Suffering Body as an Interpretative "Text"

The term of "human vulnerability" has been used in the context of bioethics recently, and sometimes it is explored in the light of human rights social justice as shown in The UNESCO Declaration on Bioethics and Human Rights, Article 8. Meanwhile, the experience of bodily vulnerability has also been connected to the phenomenology of illness in bioethical/medical discussions.

For example, in their work *Body Matters: A Phenomenology of Sickness, Disease and Illness*, James Aho and Kevin Aho attempt to surmount a new way to approach the issue of disease and illness, contending that sickness, disease, and illness are socially and historically constructed (Aho and Aho 2008). Following the core principle of phenomenology as a return "to the things themselves," they examine the phenomena of bodily afflictions and analyze them from three different perspectives: "sickness" (a sociological perspective), "disease" (a medical perspective), and "illness" (a patient's perspective). These three perspectives are definitely different from each other yet interrelated at the same time. Vulnerability therefore should be understood at these three different levels.

---

[2] It should be noted that in contemporary ethical discourse when we speak of human vulnerability we tend to contend that some people are more vulnerable than others. For instance, the UNESCO Declaration on Bioethics and Human Rights (2005) has singled out four specific vulnerable groups (children, women, the elderly, and the disabled), stating that, "human vulnerability should be should be taken into account. Individuals and groups of special vulnerable should be protected and personal integrity of such individuals respected." (The UNESCO Declaration on Bioethics and Human Rights, Article 8). In Buddhism, however, there is no such clear categorization as vulnerability is an idea applied to all human beings. As such there is no such notion as "protection" or the "rights of being protected".

By making a crucial distinction in phenomenology between two senses of the body, namely, the quantifiable, corporeal body (*Körper*) and the organic living or lived-body (*Leib*), the study by the Ahos aims to show that an understanding of the vulnerable and suffering body is an interpretative process that is more profound and complicated than biomedical notions of disease and illness. Illness then is not simply a physical phenomenon that is experienced; rather it has the same status as notions such as mind and consciousness, since it signifies the subjectivity from which a particular phenomenon is understood and appropriated. The issue of illness is also connected to the philosophical issue concerning mind/body issues. The notion of the lived-body raised by the Ahos then is a significant one, indicating a dynamic, transformative, and relational body rather than the body reduced to a mechanical entity.

Another related question has something to do with our conception of being healthy. Though the World Health Organization has a specific definition of "health." It. is true that most of us tend to experience the world through the medium of our bodies. When we are in good health, we take our healthy bodies as an ontological state of being, and illness as such is seen as "lack (or absence) of good health." In other words, illness is perceived as "other" than our personal identity, that is, our body's alienation from its primordial state of being. Traditional bio-medicine in the West hence identifies a healthy, "immune self" with an individual person who defies pains and suffering, and at the same time it seeks a method to deal with pain caused by illness. Very often illness or disease is described as an experience accompanied by underlying bodily sense of "not feeling oneself."[3] This understanding is philosophically based upon a bifurcation between self and other in which self is viewed as a fixed entity in that it seeks to maintain an identity free from "contamination" by any sense of otherness.

In recent years in the discourse of bioethics, however, some new expressions such as "existential medicine" and "medical humanities" have been used to imply an alternative way of looking at illness and health (Deckers and Gordjin 2007, pp. 357–358). These new concepts attempt to recapture an understanding of biochemical, physiological, and morphological factors in living organisms and their development in time and space, and thus offer a fresh look at health and illness. The phenomenology of the illness, health, and healing processes is seen as one that depends upon how this sense of "not feeling oneself" is interpreted and responded to. "The key question is whether the new bodily sense of self associated with a sense of dis-ease is accepted and incorporated into the individual's identity and existence or rejected by the 'immune' self,"[4] blaming either on other people or on foreign, "non-self" body tissues and cells.

The understanding of self/other in existential medicine and medical humanities to a certain degree comports with Buddhist teachings. According to Buddhism, there are three levels of suffering. At the first level, it is the ordinary suffering of

[3] 22 August 2011 The Phenomenology of the Health, Illness and Healing Processes http://existentialmedicine.wordpress.com/the-health-illness-and-healing-process.

[4] Ibid.

pain and sorrow associated with conditions we have mentioned above; at the second level, it is suffering due to our failure to understand the fact that everything changes and nothing is permanent; at the third level, it is the suffering caused by our attachment to a self that is seen as fixed, self-referential and substantive. The third one is viewed as the key reason for the suffering of the first two levels. According to the Buddhist concept of no-self (*anatman*), there is no such as thing as a self or self-substance/self-nature, nor there is such a thing as being or fixed identity.

A person thus is nothing but a unity of five aggregates (*shandhas*): They are form/body, sensation, perception, predispositions, and consciousness. It follows that what a causal relation or a causal process actually is in terms of the five *shandhas* points to a conceptual impossibility to defining personal identity. For the Buddhist, there is nothing permanently existent which can be identified as the "selfhood" over and above these five aggregates. The argument of no independently existing self is also connected to the Buddhist doctrine of dependent/conditioning origination (*pratitya-sumtpada*), the main base for the Buddhist doctrine of "emptiness" (*sunyata*). As dependent/conditioning origination indicates, self is ultimately "void" or "empty."

Yet emptiness here does not mean "non-existence" or "nihility;" rather it refers to the notion of a lack of autonomous "self-nature" or "own-being" (*svabhava*). In other words, self does not have autonomous and abiding existence since nothing, including self and self-identity, has an autonomous self-nature. What is considered as self is the result of conditions (*pratyayas*) dependent upon which things come to be, and as such one can only speak of self or self-consciousness relative to these things. Therefore, the idea of self only points to some kind of regularity of interconnectedness among events found in experience, but does not in itself spell out the logical distinctions or the degree of differences between self and other.

It should be pointed out however, although Buddhism emphasizes the notion of interconnected world and suffering as a general shared human condition or experience known to all sentient beings, it recognizes at the same time that each actual instance of suffering is different and unique. Therefore, when someone says, "I am suffering," and someone else responds by saying that "I know it," it does not mean he/she knows exactly the nature of the actual instance, but has some idea through an analogical way of thinking that enables the person to have a compassionate injunction to engage with the suffering of the other. From this perspective, Buddhism would agree to the phenomenological account that sickness, disease, and illness can be read differently due to different perspectives of the reader. Illness is a text in the sense we need to get into context and inter-texuality, or all causal conditions in a Buddhist term, in order to have a hermeneutic understanding of the text, that is, the illness itself. The acknowledgment of particularity of suffering is important, and because of this, we have to admit the limit of a universal solution of human suffering.

In the case of bioethics, we need to recognize the need of a particular person behind a disease (say, stomach cancer) that can be, but not limited to being biologically categorized, staged, and analyzed. In his essay "Hermeneutics and Experiences of the Body: The Case of Low Back Pain," *Wim Dekkers* proposes the notion of clinical medicine as a hermeneutical enterprise to bridge the gap between the general perspectives of hermeneutics and the particularities of medical practice; contending that in clinical practice one needs to pay attention to not only the biological account of illness, but also the existential account of illness, in order to fully grasp the moral dimension of human pain and suffering (Deckers 1998). Dekkers's argument aims to critique the tendency of the conceptualization of bioethics as the application of principle-based theory or the rule of natural science to clinical situations with regard to the phenomenon of pain. The critique calls for the need for practical wisdom on the part of medical professionals. It also points to an alternative form of ethics such as the Aristotelian virtue ethics that are not principle/rule-based. In terms of the Chinese tradition, the emphasis on benevolence (Confucianism) and compassion (Buddhism) in contemporary bioethical discourse attempts to do the same job.

What do we mean by saying that the suffering body is an interpretative text? A hermeneutical enterprise deals with the interpretation of texts: a literary critic looking for the meaning of a literary piece, or a judge attempting to decipher the meaning of the law. Both of them aim to have a coherent reading of their respective texts. Similarly, a physician's job is to interpret the text of a patient by relying upon diagnostic technologies to understand the possible meanings of clinical signs and symptoms. Nevertheless, the hermeneutical task performed by a physician is much more challenging since the what he/she tries to interpret is not a pure object, but a lived body with blood and flesh, and thus he/she has to face the reality that the experiential text of illness is one lived out by the patient, and that the narrative text constitutes an on-going and open process.[5]

In other words, there is always a surplus of meaning in the process of exegesis he/she has to deal with.

5.3 Compassion: The Irreducibility of Ethical Responsibilities

The physical vulnerability such as aging, sickness and death is related to the very idea that nothing is permanent (*anityā*). Suffering in this sense is an experience we all have or will have. The Mahayānic notion of compassion is intellectually and ethically rooted in the idea that all human beings are confronted with a self that is contingent and subject to vulnerability.

The word "compassion" (*karunā*) in Buddhism is usually understood as active sympathy or a willingness to bear the pain and sorrow of others. In Mahayana, compassion is one of the two qualities, along with enlightened wisdom (*prajña*), to be cultivated on the bodhisattva (i.e., Buddha-in-making) path. Chinese Buddhism translates the Sanskrit word *karunā* as *bei* (悲) which means "sympathy", "empathy", or "pity." This word is also connected to another one, "loving kindness" (Pali, *mettā;* Chinese, *ci* 慈), being rendered in English as compassion as well.

The ethics of compassion indicates two fundamental aspects in Buddhism: (1) everyone in the world is interrelated, and (2) one understands relationships in a

[5] Here I am not talking about a medical researcher working in a laboratory setting.

specific way. Unlike Confucianism where the notion of loving kindness (*ren* 仁) is characterized by familial distinctions (i.e., the graded love in a daily situation), compassion in Buddhism, however, appeals to a shared human experience, particularly the experience of human vulnerability. Compassion is an acknowledgement that when one living being suffers, we all suffer, which indicates an intense fellow-feeling one should have for all living beings who suffer pain, anxiety, ignorance, and illusion.[6] The meditative practice in Buddhism, then, involves the four "sublime state", which are (1) boundless love, (2) compassion, (3) sympathetic joy, and (4) limitless equanimity. There four states exhibit the way Buddhists cultivate their ethical life with respect to others. The bodhisattva's vow goes like this:

> All creatures are in pain, all suffer from bad and hindering karma … All that mass of pain and bad karma I take in my own body … I take upon myself the burdens of sorrow; I resolve to do so; I endure it all. I do not turn back or run away, I do not tremble … I must set them all free, must save the whole world from the forest of birth, old age, illness, and rebirth … (Strong 1995, p. 161)

The bodhisattva's vow indicates a notion of the irreducibility of ethical responsibilities towards others. Of course, not everyone can be a bodhisattva. Nevertheless, it shows the Buddhist ethical ideal. It should be noted that this ethical ideal does not come from a divine command; instead, it is based upon human experience, especially our experience of being finite and vulnerable. In a way, compassion is related to the "negative" aspects of human existence, that is, the experience of pain and suffering.

But the notion of "sharing" is a significant one since the experience of pain is no longer objectified as something "outside me", but part of "me." This notion may be at odds with the concept of professionalism, say medical professionalismin the West, which maintains a clear distinction between physician and patient, or helper and helpee. In this situation, the former is viewed as a moral agent, while the latter as a moral patient; the former has everything of value to give whereas the latter has nothing to give back (except cooperation and money). From a Buddhist perspective, however, this kind of physician/patient relationship, despite that it has been accepted as a common practice, is dichotomous, and thus lacks a real bond between the lifeworlds of two parties. What if the doctor is encouraged to have a face-to-face relationship with his/her client and vice versa? In this way, the patient is not reduced to be an object, or the subject of certain disease, but rather is a lived body, a real person. To use a Levinasian phrase, the "naked face," or the vulnerability of the other person, "speaks to us" in such a way that it cannot be reduced to sameness in terms of a perceived object.

In her heavily annotated and documented book entitled *The Meaning of Illness: A Phenomenological Account of Different Perspectives of Physician and Patient*, S. Kay Toombs provides a phenomenological account of the experience of illness

[6] Meanwhile, Mahayana emphasizes that the practice of compassion must be guided by wisdom so that one knows exactly how to treat others with a compassionate heart in a specific situations without being dogmatic. In a way, compassion arises from wisdom, and wisdom arises from compassion.

and the manner in which meaning is appropriated by the patient and the physician. Like the Ahos, she suggests that physician and patient hold different perspectives in terms of intentionality, temporality, and horizon (Toombs 1993, pp. xv–xvi).[7] When discussing the impact of a doctor's diagnosis on the patient, Toombs describe a patient's vulnerability:

> Personal unity shatters vividly at the moment one receives a definitive diagnosis of incurable illness. Suddenly all one's taken-for-granted assumptions about the world are transformed. In that instant one recognizes that nothing will ever be (or can ever be) the same again. Every patient can recall that moment in exquisite detail. It is imprinted on the mind—the marker of a transition from one way of being to another. Even though it is more than 20 years since I received my diagnosis. I can tell you what day of the week it was, what month, what clothes I was wearing, and repeat almost verbatim the words used by the neurosurgeon.

From the point of view of compassion, a mutual communication via listening is required so that a responsible and face-to-face relationship with dialogical openness can be established. A physician should not only have the medical skill to treat the disease but also the professional skill to help the patient to cope with his/her own dis-ease both physically and emotionally. According to Toombs, it is crucial for the physician to engage in dialogue with the patient in order to familiar him/herself with the patient, not only the medical symptom, but the lifeworld of the patient. Toombs does not use the word compassion, but her phenomenological approach implies the notion of compassion. The idea of compassion consists in giving comfort to others and relieving their suffering, and medical professionals, by the very nature of their job, cannot avoid doing that. This effort, in turn, requires the cultivation of kindness and benevolence (*maitriā/mettā*) on the part of medical professionals.

For most patients, the greatest fear is, perhaps, not death itself; rather the possibility that eventually the inevitable progression of their physical conditions will leave them dependent for their every need on the assistance of others, including strangers who may not care about them. The illness has created a profound sense of alienation from one's body, reinforced by self-pity and self-denial. This is the reason why Buddhist views of compassion and dependent-origination are crucial to re-orient our approach to self-other relationship as well as the self-self relationship Therefore, the key question is how we should look at the health and illness, disease and healing process, and ultimately, how we deal with human vulnerability (viz., contingency, suffering, and mortality). The ethics of compassion, similar to the ethics of care advocated by feminist ethicists, speaks more of care and responsibility in the face of human suffering. The irreducibility of our concern for the suffering other, as such, goes beyond a conceptually construed sense of justice, or an expansion of self-interest through identification with the other.

[7] Toombs also draws a distinction between illness and disease.

5.4 The Phenomenology of Illness and the Healing Process

Although Buddhism has a long history associated with the art of healing, its contribution to the medical/clinical area in the West had been ignored until the last three decades. In the essay "Buddhism in Relation to the Profession of Medicine," R. L. Soni has points out, Buddhist holistic understanding of human nature encourages psychosomatic approach to pathology of illness[8]. It is true that the modern distinction between health and illness, life and death, self and other, etc., with their sharp bifurcations gives rise to all kinds of alienation we experience in reality. Buddhism, in contrast, seeks a non-dual way to look at the body/mind issue, from which the corporeal, mental, and spiritual are seen as an organic whole.

According to the Buddhist teaching of dependent-origination, nothing stands alone apart from the matrix of everything else. Nothing is independent, and everything is interdependent upon everything else. Following this line of thinking, illness itself need not be seen as "an alien entity," but rather as something that plays an important role in the health process itself. It does so by giving psychosomatic expression to a new dimension of the self in which the "immune self," rejecting the other, (such as foreign tissues, cells, bodies, etc) may end up transforming into symptoms of illness through its immune responses Or, the disease/dis-ease can be transformed not by getting rid of it but by living with it as part of the self. As a matter of fact, the human body is continually undergoing a process of sickness and recovery. The implications of this are profound. That is to say, the otherness of the "sick self" will be allowed to coexist with the agent of sameness, so that our relationships with our bodies, healthy or sick, young or old can be more realistic and satisfactory.

Buddhism reminds us that our experience of self is never *separable* from an experience of otherness in the world. It is "the immune self" that experiences otherness as something innately separate from its own, thus creating a threat to a diluted fixed identity. What we call "health" then is not an unchanging *state*, although it is one of feeling at ease with ourselves through a familiar sense of self. Instead health is an ongoing life process, a process of becoming more "whole" which involves "becoming other." This means we have to embody the otherness of the other as new and different aspects of the self. Hence, the person who is experiencing illness may have a positive way of looking at the unfortunate situation. To use the Buddhist philosophy of the "middle way" to look at the phenomenon of illness and all the vulnerable experiences related to it, one has to reconsider the relationship between illness and health. Rather than seeing them as two opposite entities in which one has to make a choice in a gesture of either/or, one should view them as oneness that is mutually dependent and mutually transformative.

The English word "vulnerable" comes from a Latin word *vulnus*, meaning "damage", "injury" and "easily wounded." If we say healing means a process of recovery from damage, injury, or wound, illness as bodily damage, injury or wound is also

[8] Millard 1976, pp. 135–151.

part of an ongoing life process. Vulnerability in this sense also can be an opportunity to empower ourselves if we treat it with a positive attitude and realize its positive effects. Physicians have recognized that:

> [W]ithout injury, damage, and tissue breakdown, life is … not possible. Without "damaging" influences, organisms cannot develop into something new and, therefore, become vulnerable … The recently developed hygiene hypothesis, for example, states that it is precisely through a regular "damaging infectious stimulus" that the organism gains health and resistance.[9]

Meanwhile, healing is an intricate process with multiple components, which is by no means confined to physicality. In her investigation of illness, Toombs has described the loss of wholeness that is an integral aspect of incurable illness, suggesting that healing involves restoring a sense of personal harmony that does not only depend upon the integrity of the physical body but other related issues. She points out, "If healing is to occur, explicit attention must be given to aspects of illness such as bodily alienation, spatial and temporal disruption, the disorder of social relationships, and the loss of self-esteem."[10] The healing process, from this point of view, should not be confined to the physical recovery and harmony.

For a Buddhist, the cultivation of both compassion and wisdom is a matter of practice, not merely a speculative or intellectual exercise. For instance, two of the meditative practices are called *śamatha* (tranquility and stillness, *zhi* 止) and *vipaśyanā* (insight meditation, *guan* 觀), respectively. Buddhist meditative methods can be used not only as a way of relief from physical pain, but also as a way of dealing with mental disruption. While *śamatha* helps one to gain a peaceful and tranquil mind via having a fixed object of attention, *vipaśyanā* aims at self-transformation through self-observation and introspection. The former in many ways is similar to yoga and the latter as a contemplative exercise is of a higher level during which a steadfast mind becomes an active experience of change and impermanence. In other words, all these meditative practices are rooted in mindfulness, through which one's body, feelings, sensations, mind/consciousness, and objects of mind/consciousness are carefully examined. By seeing the arising and passing away of mindfulness and contemplations with regard to an origin and causality, it follows that the self related to the meditative practice arises and passes away. In a Buddhist sense, illusion or ignorance is related to the notion of permanence and having self-nature which causes a person to hold to an "I" and "mine." To see through the mode of impermanence means to examine things in transformation. By seeing their arising and cessation one can discover their impermanent nature, that is, the emptiness of all things. When talking about his own meditative experience, Jon Kabat-Zinn, author of several national bestsellers in the U.S. and a Zen practitioner, writes:

> Although at this time mindfulness meditation is most commonly taught and practiced within the context of Buddhism, its essence is universal…. Yet it is no accident that mindfulness comes out of Buddhism, which has as its overriding concerns the relief of suffering and the dispelling of illusions … I am a student of Buddhist meditation, and a devoted one,

---

[9] See the introduction to van der Bie et al. 2008.

[10] Toombs 22 August 2011.

> not because I am devoted to Buddhism per se, but because I have found its teachings and its practices to be so profound and so universally applicable, revealing and healing.[11]

Numerous clinical cases have shown that Buddhist meditative techniques are beneficial in the healing process. As Gil Fronsdal has pointed out insight mediation, or what she calls the "*vipaśyanā* movement" became popular in the U.S. beginning in the early 1980s. The meditative language adopted in the West tends to de-emphasize the sectarian differences within the Buddhist tradition, or the distinction between Theravāda and Mahayāna, since the primary goal is not about achieving a religious conversion but offering an effective method of mindfulness, stress reduction, pain management, and self-understanding.[12] Through meditation, the practitioner develops a full awareness of physical, mental, and cognitive experiences. This practice is especially helpful when applied to the process of self-healing.

5.5 Summary

The paper explicates the notion of bodily vulnerability and how illness is perceived, appropriated, and understood from the perspective of the Buddhist notions of *dukkha* (suffering) and *karunā* (compassion). It is the author's contention that health, illness and healing should be viewed as an intricate whole with multiple components, as suggested by the phenomenology of illness and health.

The discussion of the paper focuses on vulnerability as a universal human problem, which by no means suggests that the UNESCO Declaration on Bioethics and Human Rights, Article 8 is unimportant. In fact, the ideas of human rights and social justice implied in the UNESCO Declaration supplement what is lacking in Buddhism.

References

Aho, J., and K. Aho. 2008. *Body matters: A phenomenology of sickness, disease and illness*. Plymouth: Lexington.

Dekkers, W. 1998. Hermeneutics and experiences of the body: The case of low back pain. *Theoretical Medicine and Bioethics* 19 (33): 277–293.

Dekkers, W., and B. Gordijn. 2007. Medical humanities and philosophy of medicine. *Medical Health Care and Philosophy* 10 (4): 357–358.

Fronsdal, G. S. 1998. Insight meditation in the United States: Life, liberty, and the pursuit of happiness. In *The faces of Buddhism in America*, eds. C. S. Prebish and K. K. Tanaka, 163–180. Berkeley: University of California Press.

[11] Kabat-Zinn 2005, pp. 13–14, 26.

[12] Buddhist meditative techniques have been used in psychotherapy, particularly for those with depression and anxiety disorders. For the discussion on what is called the "vipaśyanā movement" in America, see Fronsdal 1998, pp. 163–80. Kenneth K.

Kabat-Zinn, J. 2005. *Coming to our senses: Healing ourselves and the world through mindfulness*, 13–14, 26. New York: Hyperion.
Millard, D. W., ed. 1976. *Religion and medicine 3*. London: SCM.
Nikaya, Digha. 1995. *The long discourses of the Buddha: A translation of Digha Nikaya*. Trans: M. Walshe. Boston: Wisdom.
Strong, J. S. 1995. *Experience of Buddhism*, 161. Belmont: Wadsworth.
Toombs, S. K. 1993. *The meaning of illness: A phenomenological account of different perspectives of physician and patient*, xv–xvi. Dordrecht: Springer.
Toombs, S. K. 2011. Healing and incurable illness. http://www.wmeades.com/id221.htm.
van der Bie, G., T. Scheffers, and C. van Tellingen. 2008. *Embracing vulnerability: reflections on healing and incurable illness*. Driebergen: Louis Bolk Instituut. Accessed 20 Nov 2012.

Chapter 6
The Ethical and the Legal Aspects of Vulnerability in the Christian Perspective

Stamatios Tzitzis

6.1 Introduction

The concept of vulnerability originates from the verb *vulnerare* which signifies "to hurt, to wound," (Rodgriguez-Arodgriguez et al. 2006). The word calls to mind a human condition made fragile by bodily or moral wounds that life's risks incur to us. These risks imply sufferings that are quasi-indelible and remain in the human person as weakening, therefore rendering the person dependent. Very often a damaged sensitivity makes critical reflection difficult and discouragement accompanied by despair isolates vulnerable individuals in a world of solitude. Let us add that vulnerability does not only concern individual persons but also indeed groups of individuals, entire communities (Weisstubet al. 2001).

Any person can encounter throughout life difficult conditions that render him or her vulnerable. Yet the idea of being vulnerable fits better the person or the group of persons belonging to a class potentially, or in actuality, socially disadvantaged, that marginalizes them in a certain manner. Thus they need attention and particular care on the part of society. Law as well as social ethics ought to provide appropriate norms to deal with this phenomenon.

Vulnerability can be an aspect of discrimination, of intimidation, or of any sort of stigma. It could also be attributed to the physique of the human body, such as the case of handicapped persons but equally to age; a typical example would be children and the elderly. As science progresses in leaps and bounds, especially in the areas of biomedicine and biotechnology, there is always a large risk of instrumentalizing the human person and in particular the vulnerable person.

(25 August 2011) Ethique et aspects légaux de la vulnérabilité dans la perspective chrétienne (trans: Letendre MC).

S. Tzitzis (✉)
Centre national de la recherche scientifique (CNRS),
University Panthéon Assas Paris II, Paris, France
e-mail: stamtzitzis@hotmail.fr

J. Tham et al. (eds.), *Religious Perspectives on Human Vulnerability in Bioethics*, Advancing Global Bioethics 2, DOI 10.1007/978-94-017-8736-9_6,

Postmodern humanism founded on the dignity of the human person requires a special attention in ethics as well as in law towards the vulnerable person. The *other*, different because he/she is vulnerable, ought to benefit from enhanced protection. The vulnerable person is more easily exposed to the dangers that provoke the reification of his/her existence.

Progress in biomedical research can, and, despite the noble intentions of researchers, has aggravated the vulnerability of certain populations subjected to experiments that intend to ameliorate the physical or mental health condition of individuals devastated by calamity. National and international legislators are responding to the call of being more and more attentive to the protection of the rights of vulnerable persons. Particularly on the international level, the European Convention on Human Rights and Biomedicine provides a most characteristic example.[1] At the French national level, legislation of March 4, 2002 and August 9, 2004 (which concern the obligation to obtain informed consent from any person submitting to biomedical treatment), plus statute numbers 2005–370 of April 22, 2005 (concerning the rights of ill persons and end of life issues) witness to the legislators' concerns for the safety and respect of personal integrity and dignity.

The great concern for the protection of vulnerable populations is inspired by the philosophy of the *Universal Declaration of Human Rights* (1948) in which personal dignity, an absolute and irreducible value, is presented as inherent to the human person, a dignity that directly joins the individual to others as being from the same human family.[2]

This declaration came into being after the shocking experience of Nazi ideology to which humanity had been subjected. The declaration affirms the same irreducible value for every human being as a person and categorically rejects the doctrine that classifies axiologically some members of humanity as superior beings and others as sub-standard beings. For the postmodern democratic world, any sort of vulnerability, even an extreme case, would not dethrone the individual from his/her ontological status as a person.

The Universal Declaration on Bioethics and Human Rights[3] draws its inspiration from the same fundamental principles. It treats the protection of persons who are particularly vulnerable by highlighting their psychic, moral and bodily integrity. Moreover, the International Bioethics Committee (IBC)[4] has concentrated its interest on article 8 of the UNESCO Declaration that concerns, first and foremost, per-

[1] For the first time at the international level, the Convention ruled on some questions which are undertaken by biomedicine in relation to human rights. The regulation date is April 4, 1997 and entered into effect on December 1, 1999.

[2] In the Preamble, we read: "Considering that the recognition of the inherent dignity of all the members of the human family and their equal and inalienable rights constitutes the foundation of liberty, justice and peace in the world."

[3] This declaration was adopted by the General Conference of UNESCO on October 19, 2005.

[4] This committee was created in 1993. It comprises 36 independent experts who oversee the progress of research in the life sciences and their applications, ever careful to safeguard the principles of human dignity and liberty. The IBC is the only world organization aimed at reflection of bioethical issues.

sons in a state of permanent vulnerability, and not only vulnerable at certain stages of their life. This constant vulnerability often characterizes individuals involved in a clinical context and as objects of new treatments derived from biomedical science and other biotechnologies. In particular, new methods and their accompanying products are tested on these individuals.

Therefore, article 8 of the Declaration, without enumerating all cases of vulnerability, intends to discern its nature which often gives way to stigmatization, marginalization and social discrimination of members of a vulnerable group, either individually or collectively.

6.2 Vulnerability and Otherness

In the postmodern world life's challenges have become very difficult; more and more they render the human person fragile above all in menacing life with a precariousness that spares no one (Bauzon 2001). They are also responsible for the development of an ethic of vulnerability strongly upheld by postmodern humanism that follows the path of modern personalism. Each one concentrates its attention on a real human and not on the abstract term of person (LOUBET DE BAYLE JL).

We are making above all reference to Levinas' humanism of the *other* which pushes man to become the hostage of a vulnerable person without conditions (Levinas 1992). Thus the destiny of this latter is placed under my responsibility, and his/her misfortunes call me and engage me before the tribunal of individual conscience and collective conscience.

But above all, vulnerability becomes an awareness of otherness, such as a suffering existence that beckons me in the name of human solidarity. This situation does not strip the fragile man of his full humanity or of his entire dignity, since distress, handicap or the inconveniences of age affecting the manner of existence ought not to devalue the ontological status of man in his quality of person. The value of man, as well as person, designates an absolute of which any attempt at proof could seriously darken the essence of all humanity. Fascism, Nazi ideology, and every discriminatory classification of human beings based on their anthropological specifications, ends up in disasters of which history has provided us with more than one example.

The ethics of vulnerability inspired by postmodern humanism does not utilize an abstract vocabulary in its discussion. Instead of citing instances of vulnerability, vulnerable persons are the heart of the discussion.

Thus all talk of helping persons in difficulty, instead of focusing on a theoretical formation of a politics of helping vulnerable persons, advocates coeducation in the life of these vulnerable persons. It becomes a practice adopted by *me* attached to the other. This translates as a change in perspective towards the vulnerable person and abolishes the distance between him and me, and invites me to approach him and make his state of vulnerability my own.

This ethics of vulnerability is then far from advocating compassion or pity for any person in difficulty.[5] It needs to understand and surround marginalized individualities, which need not be abandoned to a solitary existence, focused on sorrows and personal torments, nourished with the impression that they are different from others.

Before mentioning legal rights, the ethics of vulnerability promotes individual moral rights that separate the binary distinction—normal/abnormal—from the ontological status of the human person, making every defense for solidarity as an indispensable factor to a socially authentic life. Thus next to liberty, legality, and brotherhood as existential claims necessary to each person, vulnerability presents itself as a way of being in the human condition and implies therefore a right in its own respect; it is truly a full individual right and, in that sense, it can be lived fully and in the same way as other individual rights; even more: the three other rights would not claim their legitimately earned place in a democratic society without the recognition of the right to respect vulnerability.

If vulnerability constitutes a challenge to the nature of human existence since it reveals a tragic aspect of humanity's fragile condition, the ethics of vulnerability needs to renew at every instance its humanism into order to offer an ethic of collective responsibility with the intention of sharing the destiny of vulnerable persons as one common to all. The ethic of vulnerability inspired by postmodern humanitarianism, which supplies human rights, sees in each person a microcosm of all humanity. Since the dignity of humanity, irreducible and unquantifiable, embraces all its members in the same way, it has therefore a fundamentally democratic character. Better still the inherent dignity of one man is the realization of the inherent dignity of all humanity. In the name of this dignity, humanity becomes a welcome to the vulnerable individual in which all humanity is displayed.

Thus any norm, which inevitably concerns the well-being of humanity, takes into account the vulnerable person. The existential space of this humanity represents a field of *eros* for the other, the vulnerable person, who experiences the pleasure of living together and shares the adventures that the course of life reserves for us.

6.3 Vulnerability and *Agape*

In its own way, Christian philosophy perfects and refines this ethic in the light of its fundamental idea, *agape,* a Greek word, which remains untranslatable in every language.[6]

---

[5] It is a question of personalistic ethics based fundamentally on human rights. "We know that law aims to apply the ethical principle of justice. Consequently, the simple fact that a political view may conform to the law does not signify that it is justified on the ethical level…Alone, law does not make policies or ethical practices."

[6] Agape is a testament to the immeasurable love of God who departs from his essential attribute (immortality) to go through death in order to save humanity from death.

The ethic of *agape* goes beyond this ethic of vulnerability. *Agape* forms a cosmos in which vulnerability is the base. Here, it is not dignity, which calls for the support of vulnerable persons, but vulnerability becomes the center of this cosmos. Since the love of Christ which gives life to *agape* takes its momentum in comforting persons in distress. *Agape* goes beyond empathy (ALES-BELLO A, PALAZZINI L).[7] The former presupposes the latter and makes a call to better understand the sufferings of the vulnerable person. Since *agape* does not only imply a welcoming of the marginalized person but even more it predisposes an unconditional companionship. The humanism of *agape* is "an integral humanism."[8] The blind, the leper, the adulterous woman are all archetypes of vulnerable persons that Christ took care of in order to develop the ecumenism of his teaching.

In the Christian perspective, the necessity of helping vulnerable individuals is represented as freedom to set oneself aside in order to accomplish personhood in relationship with the vulnerable person, as being an *agapeic* relationship, the ideal of a God who reconciles himself with suffering in the name of the love of humanity.[9] Man in the image of God (from which springs his source of ontological dignity) reflects inexorably the human fragility of Christ, the divine vulnerability manifested by the Passion of Jesus of Nazareth. *Agape* is love in gifts and in demands. Love translates often the state of well-being of the one who gives to the suffering person; *agape* implies the liberation of the human person from his/her Ego in order to adopt human suffering, an imperative to relive in Christ His passion and legitimately claim the hope of salvation (Lenoir 2008). Thus, assisting the vulnerable person becomes part of the confession of faith, which engages in *agape* as the perfection of self.

In effect, Saint Paul in his *Hymn to Agape* (1 Cor. 13), instead of insisting, as is his usual manner, on faith as the indispensable element of salvation, highlights the importance of *agape* for the understanding of the world and exceeding the limits of human finiteness.

The ethic of vulnerability, in its humanistic expression, requires the integration of vulnerable persons in every social space so as to accomplish a space of all living together. This advocates the putting aside of all difference which stigmatizes marginalized persons in the name of tolerance that speaks of characteristics and weaknesses in the human condition (Durand et al. 1995).[10] The ethic of *agape* reveals a new manner of living in the fullness of existence as an existence connected to others

[7] The action of intuitive understanding of others through emotional communication which fosters an exchange of sentiments from the one speaking and the one spoken to. Empathy is attention centered on that which the vulnerable person emotionally feels.

[8] See the recent encyclical of Pope Benedict XVI Caritas in Veritate.

[9] On the thoughts of St. Gregory of Nyssa, among other remarks: "By his fault, man has abandoned the divine form… Only God can lift up the one who has fallen and bring back the lost life. God could have saved man immediately, but he has chosen the long detour of the Incarnation in order to deify him: God blends himself with our being in order that thanks to his blending with the divine, our being may become divine."

[10] On this topic, M. Andronikof remarks in characteristic fashion: "But to condemn an action, for example abortion, does not prevent the Church from welcoming those who have succumbed to this by helplessness. Rather, on the contrary, the Church is instructed to extend her mercy to them all."

to transcend his/her own weaknesses. Living actively with vulnerable persons is a sharing in their sufferings, viewed as daily situations of life, and destined to be overcome.

The ethic of vulnerability in its legal and moral expressions individualizes the case of vulnerability in order to better deal with them. In each case, it proposes aid and treatment which seem be appropriate. The ethic of *agape* universalizes; it gives access to all spaces of vulnerability as necessary spaces from which to proceed. No gift is measured by generosity nor the permission of ethical or legal norms, but in reference to Grace which asks for sacrifice. Because *agape* does not designate a simple divine love, but the sacrifice of God who out of love for man is incarnated in man and, even more: in the name of this love, he has known death, in letting go of the quality of immortality of his nature. For Christianity, vulnerability is promoted to the level of an ontological necessity of human *entelechy*.[11] Vulnerability represents thus an original status of man in the way of fulfilling his being as a person.

In the Christian perspective, vulnerability does not then allow the affirmation of an individual right at the expense of the person in difficulty. Vulnerability opens the relational space in which the suffering 'other' engages me in the life of Christ who is the affirmation of the fullness of a personal life. This point is fundamental in bioethical issues. In the name of the dignity of the person, normative rights calculate the terms of protection of vulnerable persons in order to avoid any error in scientific experiments. Notably, informed consent from a handicapped person is required for all trials in the progress of science.

To protect the life of any marginalized person, the personalistic ethic has established a sacred/secular space in which life is housed as a supreme value. It therefore fixes the object of law under the form of *a right to life*. In this context, life is subject to normative requirements of current legislation. As an individual right, it is conditioned by the cultural space of a country and submits often to its fluctuations. Then what happens is that it is counterbalanced by the ethical right to death due to the fact that an individual's life is considered to be the property of each person.

Here nihilism, in the form of the death of the suffering person, gives sense to a vulnerable existence. Countries such as the Netherlands or the state of Oregon in the USA have accepted euthanasia or assisted suicide in the name of nihilism in order to soothe suffering. Helping someone to die who is in a state of decline or one of very great irreparable suffering is considered, by the supporters of euthanasia, as compatible with personal dignity. Similar ideas reign in the ethical field of "assisted suicide" and directly concern the suicide of marginalized persons. In this particular case, stating that life is considered as a sort of individual property, each person can dispose of it according to his or her wishes. Individualism, the offshoot of the philosophy of human dignity, exacerbated, tends towards a negativistic ideology of life when one judges that his/her life is not worthy living.

This logic creates then a vision of suffering in which lesser values end up being superior to the value of the epiphany of life. Let us emphasize that the ethic of

[11] The natural end of man in relation to the accomplishment of man's being according to the inherent finality of his being.

postmodern secular personalism does not classify the fetus within the category of vulnerable persons—and therefore, does not provide a single moral norm for her protection. In effect, for this type of personalism, each individual life does not necessarily represent a personal life. Only the living newborn possesses fundamental rights. It is entirely different for Christian philosophy on the subject matter of the ethic of vulnerability. Human dignity, drawing its source from the idea of man made in the image of God, overflows into the space of *agape.*

Here, personal life is envisioned as participation in the *logos* incarnate in Christ. To be alive is a way of living and manifesting oneself as a creature in the image of God. This signifies that personal subjectivity is full of energy and moral freedom which creates relationships with others in society and binds the person before God. All moral law which initiates sharing in the destiny of vulnerable persons transforms itself into love as welcome and companionship Companionship does not signify simply helping the vulnerable person, and in particular the one who is suffering, as life comes to a close.

Companionship is a form of *sunodeia*: conjunction with the other, a co-existence in order to share the fullness of love, one for the other, in all circumstances.[12] It follows that the meaning of life does not find itself in itself but in a transcendence that draws its vigor from *agape.* So now we see that to the idea "nothing has an *a priori* meaning", the basic tenet of the ethic of euthanasia, is juxtaposed the *agapeic* affirmation of life which permeates Christian teaching.

Thus, the ethic of vulnerability in the Christian perspective, instead of reducing life to an individual right to life and explore through this lens a right to die, considers life in its irreducible and ontological dimension, incompatible with any type of nihilism in the name of suppressing suffering. Following a life centered on Christ, life constitutes an affirmation of man's being as a person made in the image of God. Life *is* and, therefore, necessarily ontological. Living in a space filled with Christ is to place oneself in the orbit of *agape.* Therefore, any trend of thought which negates the value of life as being merely the phenomenon of a personalistic transcendent deontology is incompatible with Christian teaching. All mental suffering and any and all bodily pain are the normal, secondary phenomena (*epiphenomena)* of an existence which participates in the transcendence of its subjectivity in the manner of the Passion of Christ.

In the Christian perspective, the ethic of vulnerability, instead of leading to, once the conditions of life have become extremely unbearable, a negation of life, chooses redemption as its proof. Concerning the topic of the fetus, who represents a vulnerable person in light of the incapacity of self-defense by speaking or by reacting, Christian teaching admits without equivocation the presence of a moral personality as well as a person with potential. In the name therefore of helping unconditionally every vulnerable person, Christian teaching displays hostility at any demonstration of negation of the life of vulnerable persons.

[12] συνοδεια in Greek. Plutarch. Morals, p. 891 ff.

References

Ales-Bello, A., and L. Palazzini. 2006. Empatia e implicazioni etiche. In: *La Bioetica e la Differenza di Genere,* 184–192. Roma: Studium.

Bauzon, S. 2001. *Réflexions Ethiques sur les Fins de la Nature,* 61–62. Paris: PUF.

Cook, R. J., B. M. Dickens, and M. F. Fathalla. 2005. *Santé de la Reproduction et Droits Humains: Intégrer la Medicine, l'Ethique et le Droit,* 95. Paris: Masson.

Durand, H., P. Biclet, and C. Herve. 1995. *Le médecin face au patient de religion chrétienne orthodoxe. Ethique and Pratique Médicale,* 22–25. Paris: Doin Editeurs/Assistance Publique-Hopitaux de Paris

Larchet, J. C. 1966. *La Divinsation selon Saint Maxime le Confessor,* 44. Paris: Cerf.

Lenoir, F. 2008. *Le Christ Philosophe*. Paris: Points.

Levinas, E. 1992. *De Dieu qui Vient Ã l'Idée,* 134–137. Paris: Vrin.

Levinas, E. 1996. *Humanisme et l'Autre Homme*. Paris: Biblio.

Loubet de Bayle, J. L. 1998. Le mouvement personnaliste français des années 1930 et sa posterité. *Politique et Sociétés* 17 (1–2): 219–237.

Pandele, S. 2009. *Accompagnement Éthique de la Personne en Grande Vulnerabilité*. Paris: Seli.

Rodgriguc-Arias, D., G. Moutel, and C. Herve. 2006. *Recherche Biomédicale et Populations Vulnérables*. Paris: L'Harmattan.

Weisstub, D., C. Mormont, and C. Herve. 2001. *Les Populations Vulnérables,* 6–7. Paris: L'Harmathan.

Chapter 7
Family as First Bulwark for the Vulnerable: Confucian Perspectives on the Anthropology and Ethics of Human Vulnerability

Ping Cheung Lo

7.1 Introduction

The "Proposed outline for a report on respect for human vulnerability and personal integrity" (IBC 2009), which tries to interpret the principle of human vulnerability, begins with the following observations.

Article 8 of the Universal Declaration on Bioethics and Human Rights (2005) states:

> In applying and advancing scientific knowledge, medical practice and associated technologies, human vulnerability should be taken into account. Individuals and groups of special vulnerability should be protected and the personal integrity of such individuals respected.
>
> Autonomy is a value that is essential to bioethics and its principles. Individuals behave in an autonomous way when they are capable of acting with self-determination. Medical research or medical practice cannot be conceived without considering the individual's autonomy.
>
> But not every human being is in a state or situation to execute this self-determination. These human beings are called "vulnerable" and need to be given special protection. Society has "special responsibilities" towards them. They need to be protected from all forms of exploitation or abuse, but at the same time must not be excluded from potential benefits of research. (IBC 2009, p. 2)

It seems to me that the observations above contain a certain strand of liberal individualism with emphasis on autonomy, self-determination, and the typical construal of this issue in terms of individual vs. state.[1] The drawbacks of such a construal are as follows. First, though the vulnerable individual can be protected by the overriding value of autonomy or self-determination, this emphasis permits and even encourages overly self-regarding conduct, which is not morally desirable. In a robust version of liberal individualism, one's own decision is one's "sovereign self-rule"

[1] Though this version of liberal individualism is not robust in this IBC document, an examination of full-fledged liberal, individualistic bioethics would be helpful so that we know the benchmark.

P. C. Lo (✉)
Centre for Applied Ethics, Hong Kong Baptist University,
Hong Kong, People's Republic of China
e-mail: pclo@hkbu.edu.hk

J. Tham et al. (eds.), *Religious Perspectives on Human Vulnerability in Bioethics*,
Advancing Global Bioethics 2, DOI 10.1007/978-94-017-8736-9_7,

and should be absolutely respected (Feinberg 1986); as long as the choice is one's own it should not be challenged. Second, when a person is in a vulnerable situation, the person all the more needs family members and friends around him or her to help protect him or her.

But the vision expounded above sees persons primarily as isolated individuals totally cut off from his or her web of interpersonal relationships; hence the immediate call for society to exercise "special responsibilities towards them" in the document above. Confucian ethics would not construe the issue in such terms. For Confucianism the family is the major bulwark in protecting vulnerable persons. Family plays a pivotal role between individuals and the society/state. In the next Sect. I offer a Confucian rebuttal of liberal individualism by means of its family-centric social vision. The implication of this Confucian vision for medical ethics and the ethics of protecting human vulnerability is articulated in Sect. III. Similar visions in seeing family as the first bulwark for the vulnerable in other faiths and ethical traditions are explained in Sect. IV.

The contrast between how this vision would handle a case differently and Sect. 27 of the *Report of IBC on the Principle of Respect for Human Vulnerability and Personal Integrity* is then explained in Sect. V.

7.2 The Confucian Family-centric Vision

Regarding the liberal argument of individual autonomy, some preliminary Confucian responses are as follow. First, the various significant decisions in life, *viz*., marriage, procreation, contraception, family relationships, child rearing, and education are not just personal choices; they are also familial choices. This is because the basic units of society are not atomistic individuals, but families. The Confucian social ideal is not built on mutual personal respect by sealing off part of one's life as a sovereign domain, but upon caring for and supporting one another by virtue of communal solidarity. As some scholars point out, the underpinning worldview of liberalism is individualism, and one can distinguish between ontological individualism and ethical-political individualism.

The former "involves seeing the individual as primary, as more 'real' or fundamental than human society and its institutions and structures," whereas the latter "involves attaching a higher moral value to the individual than to society or to any collective group" (Arblaster 1984, p. 15). Hence the emphasis on respecting individual autonomy. But both ontological individualism and ethical-political individualism are absent in mainstream Confucianism. In fact, John Stuart Mill's *On Liberty* was criticized by a Chinese philosopher several decades ago for neglecting that an individual or individuality is constituted by various social relationships (Hsieh 1973, pp. 85–86).[2] Tu Wei-ming, a contemporary advocate of religious Confucianism, puts it well,

[2] Within the confine of this paper the present author cannot fully articulate the Confucian social-political vision. Suffice to say that some fine scholars agree that it is unfair to characterize Confucian social thought as collectivistic. See Zhongfang 1993, pp. 321–434; Jin 1992, pp. 1–16. Some

> Confucianism conceives of the self neither as an isolated atom nor as a single, separate individuality, but as a being in relationship… Each relationship contributes to the development and overall constitution of the self. The self, in this sense, is the sum of its relationships. At the same time, it is conceived of as a centre of relationships which is not reducible to the relationships themselves…. The family is seen as an enriching and nourishing support system, a vehicle for the true realization of the self in its centre. The self, in turn, must develop in its various roles as son or daughter, parent or sibling. To do so, in each case, deepens that facet of our humanity. (Tu 1984)[3]

Second, Confucian ethics also has a hard time endorsing the link between individual autonomy and individual dignity. Since the Confucian person is, in Michael Sandel's way of putting it, encumbered and constituted rather than unencumbered and separate (Sandel 1982, pp. 53–62, 143–44), the liberal autonomy-heteronomy dichotomization is inapplicable. When one leads one's life in solidarity with one's family, such a life cannot be described as "being ushered along it by others" (Dworking 1994). The Confucian vision of the socially constituted self or self-in-relationships would not agree that "no one treats his life as having any intrinsic, objective importance unless he insists on leading that life himself" (Dworking 1994).[4] Besides, Confucian values down the ages seem to care more for family honor than for individual dignity; individual dignity just does not self-evidently trump other values.

As a familiar phrase recorded in the thirteenth century *Three Character Classic* (三 字 經 *San Zi Jing*) puts it, "Make a name for yourselves, and glorify you father and mother" (揚 名 聲'顯 父 母, *yang mingsheng, xian fumu*).[5] When one seeks honor or avoids dishonor in life, it is ultimately the effect on the family that one cares the most. Any honor or dishonor a person receives is equally an honor or dishonor for the immediate family, the extended families, and for the clan. Hence both individual autonomy and individual dignity are not *basic* values in Confucian social-political thought. [6]

---

scholars prefer to use the term "holism" rather than "collectivism" as the self-community relationship is conceived in a part-whole manner; see Munro 1985.

[3] Alasdair MacIntyre puts it aptly, "In many pre-modern, traditional societies it is through his or her membership of a variety of social groups that the individual identifies himself or herself and is identified by others. I am brother, cousin and grandson, member of this household, that village, this tribe. These are not characteristics that belong to human beings accidentally, to be stripped away in order to discover 'the real me'. They are part of my substance, defining partially at least and sometimes wholly my obligations and my duties." (Macintyre 1981).

[4] It seems that the moral equation between autonomy and dignity is prevalent only in the English-speaking world. Contemporary Germany has respecting the dignity of the human person written into the constitution, but the argument of individual autonomy is not that prevalent in that country.

[5] This saying is an abbreviation of famous passage in the first chapter of the Xiao Jing (孝 經), The Classic of Filial Piety: "When we have established our character by the practice of the (filial) course, so as to make our name famous in future ages, and thereby glorify our parents: this is the end of filial piety," King 1966, p. 466.

[6] Joseph Chan puts it well when he explains the Confucian concept of freedom, "Confucians would justify freedom only on the ground that it allows people to pursue the good. That we should be free to do X is because X is good, and not because freedom expresses or realizes personal autonomy. The Confucian justification for the freedom to do X is always content-dependent; that is, it depends on whether X is valuable," Chan 2002, p. 300.

To put things in perspective, it is important to note that the Confucian way of life is family-centered. It is common place to say that filial piety has a prominent role in Confucian ethics, but such a simple description does not go deep enough. As an example of the Confucian way of life, in a recently published article I made use of the twelve century CE neo-Confucian Great Master Zhu Xi's *Family Rituals*, especially the part "Miscellaneous Etiquette for Family Life" (*jiaju zayi* 家 居 雜 儀). Many of the materials there are derived from the chapter "Domestic Regulations" (*Nei Ze* 內 則) of the first century CE *Li Ji* (禮 記),[7] which is one of the canonical books of the *Thirteen Classics*. According to these instructions of domestic life, all married children continue to live with their parents (for males) or parents-in-law (for females) in a big family compound.

For our purpose here, there are three noteworthy features in these instructions.[8] First, decision-making of adult children is not by individual autonomy, but by permission of their parents. Second, private property is owned by families rather than by individuals. Third, marriage and divorce are also family decisions. In these three sets of instructions, one cannot find any trace of the liberal cherished value of individual autonomy. These Confucian instructions, when abused, certainly can lead and has led to parental authoritarianism in big families.

But the spirit of these instructions and innumerable similar writings are plain. The Confucian way of life is entirely family-centered and thus parents-oriented. It is my parents' life entrusted to me, not a life that belongs to me, that I am leading. Regarding what Feinberg calls "the critical life-decisions—what course of study to take, what skills and virtues to cultivate, what career to enter, whom or whether to marry, which church if any to join, whether to have children, and so on," (Feinberg 1986) these are all family decisions and the parents' wishes and preferences are crucial considerations.

In modern Chinese society big, extended families are replaced by nuclear families. But the emphasis on the family is the same. With the disappearance of dozens of close relatives living together, the urgent need for a central authority for each big household also disappears. In a nuclear family the bonding of family members is strong and many decisions can be made over the dinner table or at a bedside. A family-centered life was also parents-centered in the past, but not so in modern times for nuclear families.

Many "critical life-decisions" are family decisions in deliberation together with one's spouse and children. The rejection of the principle of individual self-determination is the same, but the principle of parent-determination is now replaced by the principle of family co-determination. Accordingly, Feinberg's ideas of individual "sovereign self-rule," the individual's "sovereign authority to govern oneself," "one is entitled to absolute control of whatever is within one's domain however trivial it may be," "the domain of his morally inviolate personal sovereignty," and of one's life "belongs to him and to no one else" (Feinberg 1986) are still utterly

[7] "Nei Ze" (內 則, domestic regulations) is Chap. 12 in the standard Chinese edition of Li Ji, but appears as Chap. 10 of Part III of Legge's translation to be quoted below.

[8] The following three observations are summary. For the textual discussion, see Lo 2010, pp. 71–72.

inconceivable in this modern Confucian family-oriented way of life. The liberal "sacred space" of individual sovereignty just does not exist in this Confucian account of the family-constituted self. For modern liberals, this Confucian way of life is heteronomy because they can think of only atomistic unencumbered individuals.

For Confucian family-constituted selves this way of life is not heteronomy because they still exercise self-determination, but this self-determination is always co-determination with family members.[9] They see no trampling of their human dignity in this way of life as human dignity should be grounded in moral authority rather than in individual autonomy.[10]

7.3 Family Co-determination as the First Bulwark Protecting Patients

Accordingly, the well-being of vulnerable persons such as children, women, the elderly, and the handicapped are all to be taken care of within the extended family, and if necessary, with the assistance of the clan. The family is the first bulwark protecting the vulnerable. Only when the family is dysfunctional or the family aggravates rather than protects the vulnerable do we need the state to step in.[11] For the rest of this article I focus on one social vulnerable group, *viz*., patients, and explore how they can be defended by the family in medical decision making.

It follows from the discussion in the last section that medical decisions are family decisions in deliberation together with one's family members. Even today, family co-determination, rather than individual patient self-determination, is the norm of

[9] This modern version of Confucian thought is my re-construction, and the mechanism of family co-determination is admittedly not neat and tidy. "Family meetings" after dinner or before bed time are usually the occasions for deliberation and decision together. There is no strict mechanism of one family member, one vote. Family co-determination of course includes self-determination, but the weight of one's voice varies in different stages of life. In general, we can say that as a small child one's voice carries some weight, but in a smaller way. Family co-determination in this stage is largely, but not entirely, parent- determination. As a teenager and young adult, one's voice in family deliberation grows much stronger. Family co-determination in this stage is largely individual self-determination. When one is married, and especially after having children, one's individual autonomy decreases again. Family co-determination is largely deliberation together with one's spouse and taking seriously the opinions of one's children. Joseph Chan also explains this well on Confucianism and personal autonomy, "Personal autonomy admits of degree—one can be more or less autonomous, and its value need not be absolute," Chan 2002, p. 301.

[10] Joseph Chan helpfully distinguishes between moral autonomy and personal autonomy and argues persuasively that classical Confucianism's concern is with the former. "The ideal of moral autonomy is that moral agents can make moral decisions that they reflectively endorse, and be able to act on this basis. Conceptually, it is possible to be morally autonomous without having valuable options concerning career, marriage, and so forth. Moral autonomy is compatible with a narrow range of life choices," Chan 2002, p. 299.

[11] Some modern Confucian thinkers do acknowledge the limit of the family, and even acknowledge sometimes family is the source of problem rather than the solution to the vulnerability of family members. I cannot go into this broad topic in this short article.

medical ethics in mainland China and Hong Kong public hospitals.[12] In obtaining informed consent, healthcare workers need to get it from key family members as well as from the patient himself or herself, though the informed consent form needs only the patient's signature. When there are disagreements between the patient and his or her key family members or among key family members themselves, medical doctors need to hold family conferences in helping them come to a consensus.

7.4 Hong Kong as Illustration

All public hospitals in Hong Kong are managed under an independent Hospital Authority. In 2002, the Hospital Authority of Hong Kong released the *HA Guidelines on Life-sustaining Treatment in the Terminally Ill*, which was authored by the Working Group on Clinical Ethics of the Hospital Authority Clinical Ethics Committee (to be referred to as "Working Group HA" below), of which the present author was a member. One salient feature of this Guideline is the pervasive language of "patient and family," "patient/family."

When the *HA Guidelines* starts with the four ethical principles of Beauchamp and Childress, there is a caveat:

> The above principles should be interpreted in the local cultural context. In the Chinese culture, the concept of self may be different from the Western concept and is more of a relational one.... The role of the family in decision-making may also be more important than that of Western societies... This document therefore acknowledges the importance of involvement of the family in the decision-making process, though the views of the family cannot override that of the mentally competent patient. (Working Group HA 2002, para. 1.3.2)

This model of decision-making is very close to what I describe as "family co-determination" above. A few more passages from this *HA Guidelines* can illustrate it better.

> The decision-making process for balancing the burdens and benefits towards the patient should be a consensus-building process between the health care team and the *patient and family*.... The health care team communicates to the *patient and the family* the realistic assessment of the patient's prognosis, i.e. the reversibility of the acute illness, the severity of underlying disease, and the expected quality of life... During such deliberations, the health care team also explores the values and wishes of the *patient and the views of the family* acting in the best interests of the patient. This fair process of deliberation and resolution, sometimes necessitating time-limited treatment trials, forms the basis for determining, and subsequently withholding or withdrawing futile care. (Working Group HA 2002, para. 4.3.3; emphasis mine)

[12] For a wonderful discussion on family co-determination in Chinese medical ethics, see Fan 1997, pp. 309–322. Fan's phrase is "family-determination," but I think "family co-determination" is a more accurate description. Fan's paper uses Chinese as well as Japanese sources for his argument. Family co-determination certainly includes the patient himself or herself. When there is disagreement within the family concerning whether or not to consent a proposed treatment, Hong Kong public hospital doctors will hold family meetings to resolve the disagreement.

In the actual practice of this model of "family co-determination" the patient is still regarded as the key factor. This is especially clear in the case of refusal of life-sustaining treatment by a competent adult patient, as the *HA Guidelines* says,

> It is good practice to involve the family in the discussion, unless it is objected by the patient, and to arrive at a consensus. However, the views of the family cannot override that of the competent adult patient. (Working Group HA 2002, para. 5.1.3)

As to the case of incompetent patients, the decision is to be made by seeking a consensus between the healthcare team and the family, as articulated in the Executive Summary of the *HA Guidelines*.

> For a mentally incapacitated patient with neither an advance directive nor a guardian, the final decision to withhold or withdraw life-sustaining treatment should be a medical decision, based on the best interests of the patient. However, the health care team should work towards a *consensus with the family* if possible, unless the view of the family is clearly contrary to the patient's best interests.
>
> To balance the burdens and benefits to the patient, the factors to consider include the effectiveness of the treatment, the likelihood of pain or suffering, the likelihood of irreversible loss of consciousness, the likelihood and extent of recovery, and the invasiveness of the treatment.
>
> Additionally, the prior wishes and values of the patient should be ascertained if possible. The above factors should be *communicated to the family to seek their views about what the patient is likely to see as beneficial,* and to aid consensus building. If possible, the decision should be taken at a pace comfortable to those involved.
>
> Sometimes, the family may not agree to a life-sustaining treatment which is considered by the health care team to be essential and for the best interests of the patient. Legally, the care team can go on with such treatment. However, other than emergency situations, a *consensus should be reached with the family* if possible. (Working Group HA 2002, p. 4; emphasis mine)

It is crucial to note that the healthcare team does not delegate the work of protecting the incapacitated, vulnerable patient entirely to the family members. As the *HA Guidelines* puts it,

> For a mentally incapacitated adult patient without a legally appointed guardian, one possible option is to apply to the Guardianship Board to appoint a guardian, especially in the following situations…
>
> - There is serious dispute among family members about withholding/withdrawing futile treatment.
> - There is evidence of wrongful motives by the family.
>
> An appropriate relative, or any other appropriate person, could be appointed as the guardian by the Guardianship Board.
>
> In case of unresolvable dispute, advice could be sought from HCE/HAHO to consider whether to apply to the Court. (Working Group HA 2002, para. 7.2.3 and 7.2.4)

The merits of this model of protecting vulnerable people in hospitals are as follows. First, patients do not have to face tough therapeutic decisions alone. In the medical consultation process a powerless patient faces a powerful physician not as an isolated individual, but as a person supported by his or her family. They are empowered by the participation of key family members, from whom they have been receiving care all along. Second, taking care of one's family members, from cradle to grave, is a much cherished virtue in a Confucian society. To participate in the therapeutic decision process of a sick family member is in continuity with one's moral duty—not a mere right—to promote the well-being of the sick family member. Third, elderly patients enjoy being taken care of by their adult children. Their overall well-being in a strange and intimidating hospital setting is enhanced when they have the peace of mind that the medical decision is a family consensus. Fourth, the patient himself or herself retains key decision power. His or her own decision will be honored by the healthcare team if the disagreement between the patient and the family members cannot be resolved. Fifth, the healthcare team is still the advocate of patients' interests. When "there is serious dispute among family members about withholding/withdrawing futile treatment" or when "there is evidence of wrongful motives by the family," the healthcare team can take the case to the clinical ethics committee and to the Court.

To revisit from this Confucian perspective the IBC 2009 Proposed outline that is quoted in the beginning of this essay, I suggest several revisions to be made (indicated by italics):

> Autonomy is a value that is essential to bioethics and its principles. *Persons* behave in an autonomous way when they are capable of acting with self-determination, *and such a self can be an expanded self that involves one's family*. Medical research or medical practice cannot be conceived without considering the *person's* autonomy.[13]
>
> But not every human being is in a state or situation to execute this self-determination. These human beings are called "vulnerable" and need to be given special protection. *First and foremost, they should be protected by their family members*. Society also has "special responsibilities" towards them. They need to be protected from all forms of exploitation or abuse, *including those by family members*, but at the same time must not be excluded from potential benefits of research.

Similar Visions in Other Faiths and Ethical Systems "A major function of the family is the care of its sick and vulnerable members" (Nelson and Nelson 2004, p. 881). This proposition is too obvious to require an elaborate defense. Hence it is not surprising at all the Confucian thesis articulated above has an overlapping consensus with many faiths and ethics, especially when the patient is incompetent.

When Patients are Incompetent It is noteworthy that one recent book on Jewish bioethics is divided into four sections: family, consent, competency, risk (Freedman 1999). The section on Family is subtitled "The role of the Family in Medical Decision Making for Incompetent Persons," and is 67 pages long. The author first articulates "the Standard View" as put forward by a number of English philosophical bioethicists—"That view holds that the usual characteristics of families—their knowledge of and concern for the interests of their members—cause families to seek, and the law to ratify, a defensible right for their participation in treatment decisions on behalf of incompetent members" (Freedman 1999). The author then

[13] To steer clear of liberal individualism, the word "individuals" should be replaced by "persons."

painstakingly analyzes this "Standard View" and pronounces its untenability for a number of reasons. The key factor in this "Standard View" is the use of the moral language of "right," and the author counterproposes that family involvement in decision making for incompetent persons should be understood as a duty rather than a right. He then goes through many Jewish legal sources and concludes as follows.

1. Children labor under strict obligations to care for their parents; the obligation of *kibud*, dutiful service; of *morah*, reverent obedience; and the strict injunctions to avoid causing their parents pain and shame.
2. Most children do not have the expertise required to perform the medical acts required by these obligations. Therefore, the usual manner in which children discharge this obligation is by appointing experts to act as their agents in providing medical care.
3. It is the prerogative of the children to fulfill their obligations by participating in these medical decisions. One who usurps this role, even when providing competent and appropriate medical care, has wronged them.
4. Some cases of medical care pose moral dilemmas; none of the treatment alternatives is ideal. In such cases, because the role of the children in medical decision making is one of strict obligation, their weighting of the relevant factors and judgment should ordinarily be conclusive.
5. The community, however, retains a general obligation to ensure that incompetent patients are not medically harmed or neglected. When the decisions of children are clearly wrong, the court is obliged to step in and order that appropriate care be provided. (Freedman 1999).

In browsing a number of Roman Catholic bioethics handbooks I have not seen a similar stress on the legitimate role of the family. But I discovered a recent article examining the suitability of the Principle of Subsidiarity for bioethics, and the author concludes that this principle "could fill a current gap in the methodology of bioethics, a discipline that has articulated a number of principles and values aimed at *how* to make a decision, yet lacks tools to address the issue of *who* should make that decision." (Kotalik 2011). Though family is not singled out as a legitimate locus of medical decision making in this essay, I see no reason why it should not be the case according to the Principle of Subsidiarity.

In English philosophical and legal writings there are considerably more articles that deal with this matter. First and foremost is Nancy Rhoden's long legal article on the two legal standards for parents' request for terminating life-sustaining treatment. She argues that on the one hand,

> the 'subjective' test, which requires the family to provide clear proof that termination of treatment is what the incompetent [patient] would have chosen, is often unworkable because a patient's character traits, and even here prior statements about medical treatment, seldom rise to the evidentiary level that courts purport to require.

On the other hand,

> the 'objective' test, which requires the family to prove that the burdens of the patient's life, measured in terms of pain and suffering, clearly and markedly outweigh its benefits, dehumanizes patients by suggesting that only their present, physical sensations count. (Rhoden 1988, p. 375)

The author then argues for a presumption in favor of family choice, proposing that a family's choice be upheld unless physicians prove that it is unreasonable. The reasonableness of a choice to stop treatment should be evaluated by considering the patient as a whole, including her values, her physical and emotional interests, and her ability to experience and enjoy life. (Rhoden 1988, p. 379). She further argues that "family members are best qualified to make these decisions, because of their knowledge of the patient's likely preferences and their special bonds with the patient… It is important to realize that the family's choice will meld the subjective and objective." (Rhoden 1988). Rhoden also understands that real life family is not without its own problems.

> But what about family abuse? Unfortunately, the same closeness that renders family members the most natural proxies may also give them the greatest conflict of interest… Yet a presumption in favor of family choice does not mean that families should have carte blanche; it means only that doctors must bear the burden of challenging the family's decisions when they appear to be based on illicit grounds… This reversal of doctor and family roles requires that the physician, rather than the family, go to court to challenge a termination decision. (Rhoden 1988)

Furthermore, "[d]isputes among family members may well require judicial resolution. Such cases should typically, however, be limited to resolving which family member is the most appropriate decisionmaker…. When no family member is able or willing to decide, as is frequently the case with elderly patients in nursing homes and mentally impaired patients" Rhoden suggests we need to have the court to appoint a non-family guardian as decisionmaker (Rhoden 1988, p. 445).

It is good to know that in the *Encyclopedia of Bioethics* there is an entry on family. The authors correctly identify the root of the problem, *viz*., personhood. "A third feature of the ethics that typifies families is a less individualistic image of persons than is customary in impersonal ethics." (Nelson and Nelson 2004).[14] This co-incidence with Confucianism is unmistakable. The authors then observes,

> Because familial relationships are not only intimate but also of long standing, family members can come to know each other in rich, particular detail and from a highly specific standpoint… This self-awareness, guided by general moral ideas such as justice, permits intimates to arrive at ethical decisions that are highly sensitive to circumstances and persons (Nelson and Nelson 2004).

When a patient is incompetent to decide about his or her own medical treatment, or when competence is intermittent, physicians turn to the family for help But in such a case both the "substituted judgment" standard and the "best interest" standard are unsatisfactory, and we need a different model.

---

[14] In an earlier article James Nelson correctly observes, "Both theoretical medical ethics as now most widely understood and medicine's own ethical tradition are ruggedly individualist: the interests of the individual patient, in splendid isolation from her social context, are to a considerable extent privileged". To be sure, there are powerful reasons for this focus, rooted in our concern about defending the vulnerability and privacy of patients. Yet there is increasing reason to believe that this intensity of focus on patient interests—considered as the interests of splendidly isolated individuals—reflects a kind of moral obtuseness, and that we would do better to design a system of medical decisionmaking sensitive to a broader range of values, Nelson 1992, p. 7.

> An ethics of the family might suggest that what family members owe each other is not the best, understood abstractly. If it were, parents would have a duty to find better parents for their children than they are themselves. Rather, what is owed is the good that inheres in this particular set of relationships. If this is right, then at the sickbed it is less important that a brother, lover, or daughter-in-law should correctly decide what is best for an incompetent patient than that the decision be made by this particular person, the one who stands as close to the patient as possible and so serves the patient as an *extended self*. (Nelson and Nelson 2004; emphasis added)

> Just as medical care is ethically inadequate when the focus is on the organ to be treated rather than on the person in whom the organ resides, so it is likely to be inadequate when no notice is taken of the families in which patients reside. An ethics that treats people as if they were unconnected and self-centered is not up to the task of promoting either justice or human flourishing. (Nelson and Nelson 2004)

7.4.1 When Patients are Competent

Much Anglophone bioethics still rests on the fundamental principle of the patient competency's, that he or she should be given complete and exclusive self-determination. Surprisingly I find an Orthodox Christian voice that resembles and even goes beyond the Confucian idea of family co-determination.

> If embedding medical decision-making within the structure of the family will lead (1) to decisions better directed twardlife and health (2) in the sense of aiding the patient's pursuit of salvation, then (3) this will be the preferable approach to patient consent. For example, if (1) there is a medical decision to be made regarding the use of (a) life-saving treatment, (b) abortion, or (c) physician-assisted suicide, and (2) if the family is more likely to guide the patient to the morally obligatory choice, then (3) they and not the patient should be the primary focus for therapeutic decision-making. (Enegelhardt and Tristram 2000, p. 364)

The justification is as follows.

> True autonomy is not capricious choice, but rightly directed choice free of the passions... Free and informed consent, when it bears on matters of salvation, should focus not just on aiding a person to choose as that person would. It should involve helping that person to choose as that person should... Free and informed consent for the Orthodox thus would not be value-neutral, nor non-directive. Nor would it be individualistic in attempting to treat the patient as an isolated decision-maker. It would instead seek to embed the patient in an Orthodox social context that can support the patient and properly direct the patient's choices. (Enegelhardt and Tristram 2000, p. 364)

In philosophical bioethics John Hardwing's "What About the Family?" (1990) remains a lonely voice. His concern arises from the harsh American reality in which healthcare is primarily private and some medical decisions can cost a family a fortune, seriously affecting other family members' interests. For working purposes he defines "family" very broadly as "those who are close to the patient" (Hardwing 1990, p. 5) and then articulates a common moral intuition about life with a family, *viz*., "to be close is to no longer have a life entirely your own to live entirely as you choose. To be part of a family is to be morally required to make decisions on the basis of thinking about what is best for all concerned, not simply what is best for yourself" (Hardwing 1990, p. 6). However, such a common morality is not reflected in American bioethics.

> Our present individualistic medical ethics is isolating and destructive. For by implicitly suggesting that patients make 'their own' treatment decisions on a self-regarding basis and supporting those who do so, such an ethics encourages each of us to see our lives as simply our own. We may yet turn ourselves into beings who are ultimately alone. (Hardwing 1990, p. 7)

It is true that this medical ethics allows room for responsible use of freedom and autonomy.

"Some patients, motivated by a deep and abiding concern for the well-being of their families, will undoubtedly consider the interests of other family members. For these patients, the interests of their family are *part* of their interests. But not all patients will feel this way." (Hardwing 1990, p. 8) Hence there is a need to modify the prevailing individual-based bioethics, "we need to consider the autonomy of all members of the family, not just the patient's autonomy… the *family* should make the treatment decision, with all competent family members whose lives well be affected participating" (Hardwing 1990, p. 9).

In this model of medical ethics, before making treatment decisions, "family conferences would often be morally *required*" (Hardwing 1990, p. 9). That is exactly what has been taking place routinely in Hong Kong public hospitals.

Unfortunately I am not aware of many publications that follow up on Hardwing's provocative arguments. His voice still needs to be heard today, as he puts it,

> The way we analyze medical treatment decisions by or for patients is plainly anomalous to the way we think about other important decisions family members make. I am a husband, a father, and still a son, and no one would argue that I should or even responsibly could decide to take a sabbatical, another job, or even a weekend trip *solely* on the basis of what I want for myself. Why should decisions about my medical treatment be different? Why should we have even *thought* that medical treatment decisions might be different?
>
> Is it because medical decisions, uniquely, involve life and death matters? Most medical decisions, however, are not matters of life and death…
>
> Have we been misled by a preoccupation with the biophysical model of disease? Perhaps it has tempted us to think of illness and hence also treatment as something that takes place *within* the body of the patient. What happens in my body does not—barring contagion—affect my wife's body, yet it usually does affect her. (Hardwing 1990, p. 6).

7.4.2 Two Illustrations

An advance directive in the form of a living will has been legally valid in Hong Kong for a few years. In 2010 the Hospital Authority of Hong Kong also produced a sample advance directive form for in-patient use. But the Authority has neither publicized it nor promoted its use. The reason is simple. A living will matters only because one wants to exercise complete and exclusive self-determination. It does not matter if we are guided by the principle of family co-determination and the anthropology of person-in-the-family in contrast to the anthropology of persons as isolated individuals. As a bioethics scholar in Hong Kong observes,

> To a connected/particularistic self, what really matters is not whether one has exercised one's rational choice (prior or counterfactual) in the last journey of life, but whether one has been taken care of in a sensitive way by one's intimate others and has gone through the journey with them together. (Chan 2004)[15]

Similarly, an American doctor reports in article entitled "Why I Don't Have a Living Will,"

> I, and surely some other patients, prefer family choice *over* the opportunity to make our own choices in advance… I have had a number of seriously ill patients say that their next of kin will attend to some choice if it comes up When challenged with the possibility that the next of kin might decide in a way that was not what the patient would have chosen, the patient would kindly calm my concern with the observation that such an error would not be very important… I believe I have a trustworthy family and a supportive circle of friends, I would prefer to endure the outcome if they "err" in predicting my preferences, or even if they choose to ignore my preferences other than the preference for family decision-making, rather than to remove from them the opportunity and the burden of making the choices. (LYNN 1991)

In short, a living will will be useful only if the first bulwark for the vulnerable does not exist or does not function in one way or another, as Lynn also acknowledges in the article (Lynn 1991).[16] When a functional family can serve as the first bulwark protecting a patient's vulnerability when the patient becomes unconscious in the last journey of life, a living will is not really important.

For the second illustration I want to go back to the publications of the International Bioethics Committee. The case put forward in Sect. 27 of the *Report of IBC on the Principle of Respect for Human Vulnerability and Personal Integrity* is worthy of our attention.

---

[15] "Similarly, an advance directive, if I really wanted to make one, should not be regarded as solely a means for me to exercise my self-determination, but should also be used to express my concern for my kin and my commitment to their well-being when I become incompetent. The directive is a means of helping them to know my voice and of facilitating the ongoing dialogue with them when I lapse into incompetency. My family members would then try to talk to me as if I were competent, but the whole point of the dialogue is not so much to figure out what I would have wanted for myself (my counterfactual choice) but to arrive at a family decision with my counterfactual participation. The prior directive only encodes my initial voice, and my voice, along with those belonging to my significant others, is likely to be transformed as the dialogue goes along. So, the final decision need not be dictated entirely by the literal meaning of my advance directive, however clear and specific it is, though it is nevertheless an important reference for my family in the decision making process." Chan 2004, pp. 96–97.

[16] Alastair Campbell argues cogently in one article that dependency is not the enemy of autonomy. For an individual ever to attain autonomy, dependency may be a vital stage in his or her progress to full autonomy. Every adult at stages in his or her life will need and desire to be dependent on others. Autonomy embraces the choice to be dependent when dependency is essential to full health and well-being. As he puts it, "But we must also accept that for some of us all of the time and for all of us some of the time the maintenance of autonomy will not be the major issue. Instead we need to know that we are responded to, loved, protected by people we can trust" Campbell 1991, p. 111.

7.4.3 Disrespect for the Patient's Will

> 27. A 78 year old man with terminal lung cancer was admitted at the emergency room because of respiratory failure. The medical exam and x-rays showed evidence of respiratory infection. The patient was transferred to the Intensive Care Unit (ICU). He asked for a no intubation order. When the doctor talked to the patient's family, to explain to them that the patient would probably need intubation, the patient's daughters challenged their father's wishes because of their religious beliefs. A few hours later, the patient had severe respiratory failure, but he refused to be intubated. Minutes later he fell into a coma, was intubated and connected to mechanical ventilation. The next day he underwent dialysis. During the next 20 days, he continued to be intubated, underwent dialysis, received antibiotics, mechanical ventilation and hemodynamic support, until he finally died.
>
> *The nature of the vulnerability*: The clear wishes of the patient were disregarded and his autonomy was thus disrespected. There is no suggestion in the case that his autonomy was impaired at the time he made his wishes clear.
>
> *The cause or context of the vulnerability*: The principle of informed consent is at risk whenever someone claims to know what is the right thing to do, and insists that his or her decision should prevail over the self-determination of the patient, whether that person is the physician or a family member. In this case, the precarious condition of the patient alone cannot justify overriding his wishes but the dire consequences of not intubating and the pressure from the family serve to undermine respect for patient autonomy.
>
> *Remedies*: Reinforcement of the need to protect an individual patient's integrity, including specifically the importance of respecting the right to refuse treatment. Appropriate clarification of the legal relevance of the views of relatives of incompetent patients should be provided to healthcare professionals as part of their professional education. (IBC 2011, p. 7)

From the Confucian perspective articulated above, which has overlapping consensus with some other ethical systems, the suggested remedies of the above case are too simplistic and should be modified as follows.

1. Before our case begins, prior to admitting into the emergency room, the family members of this 78 year old man should know that he has terminal lung cancer. They should have initiated a family discussion on the various options in advance care planning. Actually the doctor who made this diagnosis should have initiated a family conference on advance care planning for the patient.[17] In short, long before the patient's admission into emergency room and ICU, a detailed discussion on whether or not to forgo some life-sustaining treatments should have taken place. If the patient has thought about refusing artificial ventilation he should have informed his family members, and they should have expressed their views and had some genuine discussions. They might have a consensus back then, or they might not.
2. When this case unfolds in the way it is described above, the patient's daughters should talk with their father upon knowing his refusal of artificial ventilation.

[17] The Hospital Authority of Hong Kong understands advance care planning as "a process of communication among patients, their health care providers, their families, and important others regarding the kind of care that will be considered appropriate when the patient cannot make decisions."

While they can persuade him to withdraw his instruction to the healthcare team, they should also listen to his concerns. The daughters should try to understand their father's serious plight toward the end of life's journey, offer emotional support at bedside, and concede that it is their father who is suffering spiritually as well as physically. They should admit that their father is making a decision wholly unlike other decisions in his life in spite of his fatherly responsibilities, and so his informed choice should be respected with utter seriousness. It is understandable that a father in such a frail and dying condition usually makes medical decisions in self-regarding terms.

After all, it is his own body that is disintegrating and it is he who is feeling the psychophysical suffering.[18] No one should charge him for being selfish when he decides to forgo life-sustaining treatment. Besides, religious belief cannot be forced upon another person, even if that person is one's father.

3. As a father habitually does, however, he should also think about the impact of his refusal of artificial ventilation to all his immediate family members. He should communicate clearly his wish to them and help them understand that he is not making an impulsive choice.
4. The doctor should initiate and facilitate this family conference before and during hospital stay as a part of advance care planning for the patient, trying to forge a family consensus.
5. "Some family conferences about treatment decisions would be characterized throughout by deep affection, mutual understanding, and abiding concern for the interests of others. Other conferences might begin in an atmosphere charged with antagonism, suspicion, and hostility but move toward greater understanding, reconciliation, and harmony within the family. Such conferences would be significant goods in themselves, as well as means to ethically better treatment decisions. They would leave all family members better able to go on with their lives" (Hardwing 1990, p. 10).
6. Ultimately if the disagreement between the patient and her daughters cannot be resolved, the doctor should talk to the clinical ethics committee, and go to the court if time permits. If the case is too urgent to be reviewed elsewhere, the doctor should honour the patient's autonomous refusal throughout rather than honouring it when he is conscious and succumbing to the daughters' demand when he falls into a coma, as the case describes.
7. The scenario recorded in the IBC Report above is an instance of the model of family-determination, which unfortunately happens in some places of the world. In contrast, the model of family co-determination, which recognizes the patient's key voice, and as practiced in Hong Kong routinely, would ask the doctor to do his or her best to forge a family consensus ahead of time, and would honour the patient's autonomous refusal if the consensus is not forthcoming.

[18] I want to thank the anonymous reviewer for pointing this out to me.

7.5 Conclusion

A finding of this article is that some public statements of the International Bioethics Committee are rather Western-leaning in moral reasoning. It is the hope of the present author that the presumption for individual-based medical decision making can be balanced by the presumption for family-based medical decision making in the future deliberation of the International Bioethics Committee.

Bibliography

Arblaster, A. 1984. *The rise and decline of western liberalism*. Oxford: Basil Blackwell.

Brazier, M., and M. Lobjoit. 1991. *Protecting the vulnerable: Autonomy and consent in health care*. London: Routledge.

Campbell, F. 1991. Dependency revisited: The limits of autonomy in medical ethics. In *Protecting the vulnerable: Autonomy and consent in health care,* eds. M. Brazier and M. Lobjoit, 101–112. London: Routledge.

Chan, J. 2002. Moral autonomy, civil liberties, and confucianism. *Philosophy East and West* 52 (3): 281–310.

Chan, M. M. 2004. Sharing death and dying: Advance directives, autonomy and the family. *Bioethics* 18 (2): 87–103.

Dworking, R. 1994. *Life's dominion*. New York: Vintage Books.

Enegelhardt, J. r., and H. Tristram. 2000. *The foundations of Christian bioethics*. Lisse: Swets & Zeitlinger.

Fan, R. 1997. Self-determination vs. family-determination: Two incommensurable principles of autonomy. *Bioethics* 11 (3–4): 309–322.

Feinberg, J. 1986. *Harm to self*. New York: Oxford University Press.

Freedman, B. 1999. *Duty and healing: Foundations of a Jewish bioethics*. New York: Routledge.

Hardwing, J. 1990. What about the family? *The Hastings Center Report* 20 (2): 5–10.

Hsieh, Y. 謝幼偉. 1973. A critique of Mill's "On Liberty" ("穆勒"論自由"的批判). (Papers in Chinese philosophy (中國哲學論文集)). Taipei: Hua Kang Publishers.

International Bioethics Committee (IBC). 2009. *Working document on the principle of respect for human vulnerability and personal integrity*. Paris: United Nations Educational, Scientific and Cultural Organization.

International Bioethics Committee (IBC). 2011. *Report of IBC on the principle of respect for human vulnerability and personal integrity*. Paris: United Nations Educational, Scientific and Cultural Organization.

Jin, Y. 金耀基. 1992. Individual and group in confucian thought (Rujiao *xueshuo zhong de geti he qunti*, 儒教學說中的個體和群體). (Chinese society and culture (中國社會與文化)). Hong Kong: Oxford University Press.

King, Hsiao. 1966. *The sacred books of China: The texts of confucianism* (trans: Legge, J.). vol. 3, ed. F. M. Müller. Delhi: Motilal Banarsidass.

Kotalik, J. 2011. Examining the suitability of the principle of subsidiarity for bioethics. *Kennedy Institute of Ethics Journal* 20 (4): 371–390.

Lo, P. C. 2010. Euthanasia and assisted suicide from confucian moral perspectives. *Dao: A Journal of Comparative Philosophy* 9 (1): 53–77.

Lynn, J. 1991. Why I don't have a living will. *Journal of Law, Medicine & Health Care* 19 (1–2): 101–104.

Macintyre, A. 1981. *After virtue: A study in moral theory*. Notre Dame: University of Notre Dame Press.

Munro, D. J. 1985. *Individualism and holism: Studies in confucian and Taoist values*. Ann Arbor: University of Michigan Press.

Nelson, J. Lindemann. 1992. Taking families seriously. *The Hastings Center Report* 22 (4): 6–12.

Nelson, H. L., and J. Lindemann Nelson. 2004. Family and family medicine. In *Encyclopedia of bioethics*. 3rd ed., ed. S. G. Post. New York: Thomson-Gale.

Rhoden, N. K. 1988. Litigating life and death. *Harvard Law Review* 102 (2): 375–446.

Sandel, M. J. 1982. *Liberalism and the limits of justice*. Cambridge: Cambridge University Press.

Tu, W. M. 1984. *Confucian ethics today: The Singapore challenge*. Singapore: Curriculum Development Institute of Singapore and Federal Publications.

Working Group on Clinical Ethics of the Hospital Authority Clinical Ethics Committee (Working Group HA). 2002. HA guidelines on life-sustaining treatment in the terminally Ill. Hong Kong. http://www.ha.org.hk/haho/ho/cc/clinicalethicreport_eng_graphic.pdf. Accessed on 14 Sept 2011.

Zhongfang, Y. 楊 中芳. 1993. Are chinese really "Collectivists"? (中國人真是'集體主義'的嗎). In *Chinese values: Perspectives from social sciences* (中國人的價值觀社會科學觀點), ed. Y. Kuo-shu (楊國樞). Taipei: Wreath Publishers.

Chapter 8
Between Tradition and Modernity: Bioethics, Human Vulnerability and Social Change

Prakash Desai

8.1 The Hindu Tradition

The Hindu civilization has been a continuous tradition for at least the last 3,500 years, and it has also always been in transition. The Indians think of their tradition as a stream, *pranalika,* flowing at various speeds and changing with gradients, adapting to local conditions. The classical metaphor invoked by Jawaharlal Nehru (1981) for this tradition was that of a palimpsest, an ancient parchment upon which generation upon generation has written its message, without quite completely erasing the previous ones. To invoke another metaphor, it is like an archeologist who comes upon strata of artifacts from different ages, in India one comes upon different eras simultaneously. On a modern city street in India the scene of a bullock cart/wagon transporting a large drum of gasoline captures this strange mixture of time periods.

When viewed from afar the Indian civilization or tradition may seem static or stagnant, even hide-bound, but change it does, ever so slowly and rarely through a radical departure, among a people who greatly value their sense of rootedness in their history. The pride with which Indians say "we are an ancient civilization" bespeaks of the respect in which they hold the past. They look to the ancient tradition for inspiration and guidance, but are always open to subtle adaptations.

An important feature of Indian tradition is that it is non-canonical. In offering a preface to my book "Health and Medicine in the Hindu Tradition" (1989), Professor Martin Marty, the celebrated Protestant theologian and historian exclaimed "where is the canon?" In fact there are no fixed, abiding commandments that all Hindus must follow. The Hindu story is one of slow movement away from a center; through interaction with local traditions and circumstances making locally appropriate and accommodative changes in practice. There is no central authority, no law-giving church, no priesthood that is the final arbiter or interpreter, no single book and no messiah. Wherever you look diversity is the rule.

P. Desai (✉)
The University of Illinois, Chicago, USA
e-mail: PDesai@psych.uic.edu

J. Tham et al. (eds.), *Religious Perspectives on Human Vulnerability in Bioethics,* Advancing Global Bioethics 2, DOI 10.1007/978-94-017-8736-9_8,

This adaptability is best understood by examining the more recent Indian history, especially the interaction with both religions and socio-political ideologies and institutions of the West. The British raj that prevailed in India for over 150 years is an interesting case study of this social evolution, and demonstrates another feature of traditions. Traditions generally and by themselves are slow to change, but change they do. History bears out an observation that sometimes it takes looking outside of one's tradition to make a significant leap out of a closed system.

8.2 Religion

Hinduism speaks in many voices and has many centers. Since no one God became preeminent, monotheism did not prevail and a singular explanation of the origin or coherence of life never became a dogma throughout India. Moreover, unlike Greece and the Middle East, neither did patriarchy prevail. Instead, what emerged was a comingling of the Indo-European patriarchy and the native Indian matriarchy, and again true to the diversity of all things Indian, the proportions of this comingling varied from region to region in India. Thus multiple gods and goddesses were worshipped, each a dominant force within a particular sphere of activity, locale, and time frame. The religion itself remained, in theology as well as liturgy, nondogmatic and noncanonical, and no central organizing institution emerged to make up for the absence of a single prophet or a unifying text. Hindu ethos and praxis are variable in the same locale and penetrate local traditions with subtle variations.

Adding to the complexity and variety in the religions of India are two other old religions that mark the period of transition around 500 B.C.E. when the *Upanishadic* revolt within the older Vedic sacrificial tradition reached its climax. Buddhism and Jainism arose in India around the same time. Gautama Buddha (Buddhism) and Mahavira (Jainism) were princes and, as non-Brahmins were in a position to more easily challenge ritualistic religion involving animal sacrifice and the attendant social order that divided people into a hierarchy. Both religions championed the cause of ahimsa (nonviolence), and both rejected the notion of social class. They sought an egalitarian society in which people were taught the path to salvation without the ministrations of Brahmin priests who wielded the authority of the proto-Hindu scriptures, the Vedas. Both preached an ascetic ethic through a life of austerity, avoidance of pleasure seeking, and the courting of suffering through reduction of one's bodily needs.

Jainism emphasized the many-sidedness of truth, that no one point of view of approach allowed a full appreciation of the truth, that in fact absolute truth or knowledge was impossible, and hence there was no place for dogma. In matters of conduct, the Jain faith extolled five virtues (1) nonviolence, (2) truth-speaking, (3) not stealing, (4) chastity, and (5) nonattachment to worldly things. Mahatma Gandhi derived many of his ideals from these tenets.

Buddhism had a profound impact on Indian history. This atheistic philosophy regards suffering as central to life; the fact of death annihilates the value of the pleasures of life. Buddhism views everything as caught in a web of transformation. Since ignorance veils the true reality, once one is removed from the unknowing state, one can enter Nirvana (literally the blowing out of air, or extinction of a candle). In its ethics, Buddhism conceived of compassion and nonviolence as the heart of its teachings. Both of these heterodox systems retained the Hindu theory of karma, the doctrine that every action generates automatic and inexorable consequences.

At later stages Islam, Christianity, and Sikhism spread in India, following the gradients of caste and regional differences. Christianity had spread to India from the early centuries of the Christian era mainly through missionary work but later hastened by identification with the religion of the ruling class. Islam also spread through the efforts of missionary Muslim clergy, but forced conversion by the ruling class was a significant contribution to massive change. On the other hand, the spread of the Sikhism in the fifteenth century and later was essentially regional, mainly in the Punjab, an area that had been militarized by centuries of foreign incursions through the Northwest frontier. Each of these three faiths offered dignity and equality to all under a common umbrella.

Another facet of our concern is the Hindu medical tradition, *Ayurveda*, the ancient medical enterprise. *Ayurveda*, the science (knowledge and understanding) of life, was an early development in Indian history. Based on a variety of Indian philosophies (like Yoga, *Samkhya, Vaisheshika*, etc.) and in tandem with religious understanding of the body and its constituents as well as the mind and its various faculties, a medical enterprise grew into a systematic and organized medical theory and practice. Primary sources of this ancient system were the compendia known as the *Carakasamhita* and the *Sushrutasamhita,* the first dating to between the first century BCE and the first century CE whereas the second dating to the fourth century CE. These texts also formulated principles of medical ethics. Devotion to the profession out of empathy for the living rather than personal gain, devotion to continuous learning, not prolonging suffering, respecting privacy, etc. were some of their principles. Consulting the family in all matters of health was seen as of paramount importance.

8.3 Social Divisions in the Hindu Culture

The division of the social order into four classes is perhaps the most distinguishing and most notorious feature of the Hindu tradition. According to the *Rig Veda* (X-90), the four classes were created from the body of the primeval person, a cosmic giant (*Purusha*). In actuality, skin color played a part in this division since the immigrating/invading Indo-Europeans were white and unlike the original inhabitants, the Harappans, the people of the Indus and other tribals. The invaders achieved military superiority over the natives, regarding them first as slaves or slave-like, and later accorded them a lower rank in the social order.

By the time of the literature of the Law books (*Dharmashastras*), the four normative classes of *Brahmins* (priests, scholars, teachers who knew and had mastered the sacred lore), *Kshatriyas* (rulers and warriors), *Vaishyas* (traders, bankers, agriculturalists), and *Shudras* (toilers and menial workers) had become fixed. This original division of the society appears to have provided a schema for a division of work; the lowest designation was conferred on the dark-skinned natives who were consigned to do manual and menial labor. Untouchability, a later development, derived from states of permanent pollution the members of the lowest class suffered from contact with such substances as dead bodies, human waste, and animal products. This contact placed such polluted people outside the system. For Mahatma Gandhi, this was by far the most shameful aspect of the Indian tradition, and he worked mightily to reform the system.

The differentiation into classes (*varna*) was further accented by guna theory, by which certain qualities (*gunas*, literally "strands") inherent in particular castes were further accentuated or altered by occupational tasks, foods ingested, activity undertaken, transactions entered into, etc. Three *gunas* are found in all material substances; goodness (*sattva*), vitality (*rajas*), and inertia or darkness (*tamas*). In the four-fold *Varna* (class) system, the Brahmins are regarded as *sattva* dominant, the Kshatriyas as *rajas* dominant; the vaishyas are regarded as having mixed dominance; and the Shudras are regarded as *tamas* dominant. In the development of Hindu thought, a person's dharma was derived from his class and the accompanying *gunas*. When the religious text *Bhagavad Gita* declared that death in pursuit of one's *dharma* was preferable to following someone else's, it invoked the guna theory of the inherent differences between classes of individuals, another feature that accorded higher or lower status.

The modern division of Indian society into castes is often erroneously thought of as identical with these four classes, but it is not. The Indian word *jati,* or caste, is cognate with Latin *genera*, and similarly denotes a group of people sharing common characteristics, for instance occupation and ancestry. *Jatis* are governed by the rules of endogamy (marrying within the limits of one's caste), and commensality (eating together). Strict adherence to rules of marriage on the part of the group preserved the boundaries of the caste, and rules about exchanges of cooked food reinforced notions of purity and hierarchy between castes, thereby stressing the inherent "nature" of a person belonging to a particular caste. A peculiarity of the system was that the caste system conferred a place in hierarchy, usually rigid and fixed by birth, but allowing some upward or downward mobility by virtue of special achievements or failures, rarely by an individual but usually involving the whole *jati.*

From any vantage point the situation of the Untouchables, the suppressed and backward "classes," was the most precarious. There was no dignity in being an Untouchable, neither was there redemption. Thus the Untouchables and the tribals were natural targets for the Christian missionaries in their efforts at conversion, and many Indians found in the new religion not only help for their social and economic state but an alternative for a life of indignity. It was natural, then, that the Hindu reformists were keenly attuned to the problem of the "backward classes" in particular and the caste in general. Mahatma Gandhi's interest in diminishing the prejudice against the *Harijans* (the Untouchables), the Children of God as he designated them, may be one of his most lasting contributions to Indian society.

As the ancient law givers tried to organize the social order, they introduced a system of hierarchies between classes of people, and also along lines of age and gender, that led to iniquitous distribution of power and skewed interpersonal transactions through domination and exploitation. The vulnerabilities of different populations, of different genders and ages, are thus systemically built into the order and have, over the centuries, emerged as the soft under belly of the Indian civilization.

Two other groups of people stand out as vulnerable to many forms of inequities and discrimination. First are the so called tribals, or the scheduled tribes (as enumerated in a schedule of the constitution) who have lived on the periphery of the Hindu society, away from established civilization, mostly in the forest, who have only until recently been food gatherers and have been outside the Hindu religious fold, that is not yet "Sanskritised" by the mainstream of Hindu tradition. The second group is that of the religious minorities of Muslims and Christians. Many from among these groups, unevenly spread all over India, are religious converts from the backward castes some generations ago, that is the untouchables and the scheduled tribes who carry a stigma to begin with. The prejudice against either of these two groups, though not recent, has hardened and intensified. Muslims have become suspect in terms of their loyalty to the Indian state since the partition of India into mostly Hindu and mostly Muslim states, and both Christians and Muslims have become targets ever since the rise of a virulent form of Hindu fundamentalism.

Conversion from Hinduism to either of these faiths is experienced as an attack on Hindu pride. The converts may in fact have found a life of religious dignity and their conversion may have led to improvement in their living condition, but by and large their lot has been poor. At the same time there have been social movements to address these vulnerabilities. In about the last 200 years the pace of these reform movements has quickened, especially in face of European missions and in emulation of them.

8.4 Sources of Strength and Protection

Before reviewing the pattern of social change in the Hindu tradition in more recent times, we must examine the sources of strength and vulnerability intrinsic to the tradition. The Hindu kinship organization provides protection to persons in the network and lends cohesion to the organization. This kinship organization is marked by resilient connectedness (a part of the interrelationship of all things). In creating order in the social organization not only is hierarchy maintained but also a system of mutual obligations, responsibility, and loyalty, making the network resilient in the face of life stresses.

This strong sense of solidarity protects all stages of dependence. Mutual obligation and support become values that take the place of Western individual autonomy. In the west, particularly in the U.S., the negativity associated with dependence has increased the vulnerability of the sick and disabled, who are experienced as parasitic. The elderly and the poor, the disabled and dependent—in other words, all needy people—are seen as unproductive. In the Indian world in general, and the Hindu in particular, the elderly receive a great deal of respect and deference in virtue of their higher status in hierarchy of age and become themselves the protectors of the young, women, and children.

The weak and the disabled are seen as deserving protection. The otherwise unproductive have a right, and are entitled to support and succor. Given that connectedness is so highly prized, all unnecessary discord within the family is avoided and that which cannot be avoided is easily glossed over. Collaboration, or at least getting along as opposed to competitiveness, is promoted from childhood onwards. Family becomes a strong nexus bringing all under its fold and taming all manner of individual interests, especially those that are mutually exclusive or harmful to cohesiveness, and thus promotes a vision of welfare of the unit over that of the individual.

Family is a rich social support network that boosts individual resilience against disease also protects against natural calamities and disasters. On a larger scale, caste networks perform a similar function although in a somewhat less personal way. The Hindu tradition of philanthropy tends to focus on caste networks, but also provides for donations to the poor, the sick, the socially dependent, and the needy. Compassion and mercy shape these attitudes and activities.

Charitable trusts create and operate clinics, hospitals, dormitories for students and orphanages that cater to particular caste networks, but are based on more general Hindu principles. *Caraksamhita,* the oldest Sanskrit medical text, declares that the motivation to pursue the medical profession must not stem from any form of self-gratification, but only from compassion towards all living organisms.

Social reform movements that eased transition into a successively more modern world derived their inspiration not only from Western ideas of liberalism but also from native and traditional Hindu and Buddhist attitudes towards suffering, particularly compassion and mercy. It also is an aim of *dharma* to protect the weak and the helpless and is obligatory for a ruler.

8.4.1 Sources of Vulnerabilities

Social and the religious forces and strivings in India are confluent and hard to tease out. Nevertheless, the organization of society and the place and tasks of its members clearly derive their sanction from religion.

Dharma, or the "Law" governing adherence to the pursuit of virtue, is the overarching principle of Hindu life, and within it are contained the additional two goals of life: *artha*, or pursuit of a means of livelihood, and *kama*, the quest for pleasure. For each person, the tasks of adhering to virtuous conduct throughout life, the acquisition of a means of livelihood, and the seeking of pleasure, are determined by the caste one is born into; status, age, and gender give rise to hierarchies of power, access, etc. These constraints apply not only to behaviors in the religious sphere, vis-à-vis God, but also to relationships with others, including husband-wife, parent-child, teacher-pupil etc. Each relationship mandates mutual obligations according to the Law. These are hierarchical and not reciprocal, hence vulnerable to exploitation and oppression because of the status and, in practice, a power gradient.

The fourth aim of life, *Moksha*, liberation or deliverance, is in a different category from the other three. Although release from the bondage of work (karma) and the cycles of birth and death is encompassed within the idea of the Law (Radhakrishnan and Moore 1973), *Moksha* is essentially otherworldly, and so forms an end in itself.

The status of women in the Hindu culture was subordinate to that of men, although a woman is essential for the completion of a man and the discharge of a man's debts to Gods, sages, and forefathers via her role in the acquisition of progeny (particularly sons). Women, in general, were seen as undeserving of Vedic learning, respect, or admiration. As wives they were regarded as inferior, easily given to sexual excitement, and the draining of a man's energies. They were said to be prone to lapses in virtue, thus causing strife in the relationships within the family. As mothers, however, they were exalted, instruments of generational continuity, nurturing, and thus entitled to respect for their maternal functions.

Fear of the power of the sexuality of women was a prominent factor in the practice of prepubertal marriages and in the isolation of women within the home. The early social reform movements focused on these inequities and cruelties against women, from the custom of urging self-immolation for a widow on the funeral pyre of her dead husband to the practice of child marriages and physical abuse of women by their "entitled" husbands. As India modernizes, as we shall see later in the paper, contact with the western world and its values of individual freedom and equality have had a singular impact on the status of women. In post-independence India laws have been enacted to protect women, for example by fixing a minimum age of marriage to prevent child marriages, giving women rights to parental property equal to that of sons, and more recently legal recourse for women against demands for dowry, as well as special provisions to investigate suspicious deaths, the so called dowry deaths. A vast change has occurred as more and more women are educated, especially at colleges and universities, and then themselves act as vigilant observers and social change advocates.

The status of children, another possible vulnerable group, divides along gender lines. Sons are greatly valued as sources of generational continuity and keepers of faith. Among the debts a person has debt to one's forefathers is a primary one. In life sons support their parents, not leaving the parental home at maturity or marriage, but rather becoming part of a joint family system sharing the hearth and the purse in various combinations; especially in traditional India, both rural and urban. After death a son performs important funerary rites, and annually thereafter on the anniversary of the death, to ensure safe passage to the world of the forefathers and general well-being while there.

Daughters on the other hand are a burden. Not only do they leave the parental home to go reside with the husband's but are a source of anxiety when unmarried, and usually require a large dowry at marriage, often beyond the means of an ordinary family. They are a mixed blessing, both a source of joy in most families and of great misgivings as well. These attitudes result in very disparate treatment of male and female children as far as nutrition and medical care are concerned. Given high infant mortality rates female babies bear the brunt of neglect.

In a time not so long ago female infanticide was known to occur not infrequently, especially in the north-west and Hindi speaking areas of the country. In our contemporary times, abortion having been legalized in 1972 without much religious protest, the more recent advances in reproductive technology, viz. pre-natal gender diagnoses has resulted in disproportionate female feticide. These practices, that is infanticide, female feticide and neglect of female babies, have led to a high masculinity ratio (Singh 2010). For example the total sex ratio for India is 940 females to 1,000 males, a reversal of trends in the west. Pointing up regional differences are the figures of 877 females to 1,000 males in Haryana, a northwestern state and 1,084 females to 1,000 males in the southern state of Kerala (a very complex story to go into in any detail about here) for the year 2010 census.[1] The selective abortion of female fetuses reached such scandalous levels that several states have enacted laws that prohibit pre-natal diagnostic tests for gender identification, both the physician or the laboratory performing the test, and the couple seeking it, punishable by law.

The gender divide does not operate with the elderly. In a society that values hierarchies of all manners, old age is a matter of veneration. As we saw earlier, households tend to be joint, i.e. several generations residing together as the older parents assume the mantle of leadership, whose consent in all matters pertaining to the young is essential. The *pater familias* presides over all important occasions, religious and secular, dispenses resources, makes decisions and generally directs the life in a household. Older women, although not to the same extent as older men, exert powerful influence and are deferred to in most matters. Older parents' status lends cohesion to the entire family, and is a source of strength and of solidarity. In as much as they are entitled to deference and even obedience, they have a reciprocal obligation to extend patronage and provide protection, guidance and emotional bonding.

It is here that the principle of family solidarity comes into conflict with the ideal of autonomy. Even in the ancient Hindu philosophy, a natural tension was visualized between autonomy and connectedness. We may translate this *Upanishadic* vision of the tension thus: Alone one cannot play but fear comes from the second. Family obligations can become oppressive, the power gradient can lead to abuse, and the desire for autonomy may get suffocated. Under these circumstances one has to find ways to escape. A young man who, not yet of the age at which one resigns from active engagement with the social network and retires to the woods, decides to leave home to don the saffron, was certainly a more traditional way out of the family's web. It was a kind of assertion of one's separateness from others. But in more recent times, young men and women have simply chosen to march to their inner drummer and abandoned the cloak of solidarity altogether. Some, especially women, are not able to escape, and for them there is continual suffering.

[1] India Guide 2012.

8.5 Social and Political Change

In order to appreciate how a tradition changes and modernizes itself we'll now review the changes brought about by the coming of Europeans to India, first to trade and then assuming political power and control, establishing their rule. The reason for review of these changes is to appreciate that faith traditions tend to be closed systems, mostly impervious to change, and, if there is some openness, this often requires influences from the outside, from another tradition or ideology with a different value system.

The era of colonial occupation in India by European powers began with the landing of Vasco da Gama on the southwestern coast of India in 1498. Da Gama had come looking only for "Christians and Spices" (SPEAR 1978), but he inadvertently opened the floodgates of the Indian markets not only to the Portuguese, but also to the Dutch, French, and English merchants who established a political foothold in different parts of the country over the next century.

It cannot be overemphasized that, save for short periods of military conquests, India never was a cohesive society or nation. There was a certain geographic and historical identity, but even with regard to religion it was not a homogeneous unit. Fragmented into a myriad of small and big states politically, and into rigid hierarchic order socially, people as a whole held diverse perspectives and their aims were sometimes mutually antagonistic. Political and social exploitation was the rule in a highly stratified society, where most of the privileges were reserved for a small minority of the political and cultural elite. To a vast number of people making only a subsistence living, with any number of masters, neither the British nor the Moguls before them were different from other oppressors.

By the end of the eighteenth century the British successfully established their rule over large parts of India; they subdued local uprisings, extended protection to local princes in return for their hegemony, and achieved total military superiority. By the end of the eighteenth century, the British East India Company had conquered most of India. The company brought the entire country under central control and a single administration for the first time in Indian history.

8.5.1 Reform Movements

When a new era began with the administrative relationship between Western rulers and their elite Eastern subjects, a more direct collision between the two cultures also erupted. The European religious establishment began to view their mission in India as one of converting the heathen to the "Word."

The chronicles of the missionaries described Indian natives as people who were savage, primitive, heathen, and debauched. Both in India and in England, the British elite took upon itself the task of civilizing India (Hutchins 1967). The East India Company in its turn began to redefine its mission as that of "introducing" Western ideas and institutions. To dispense justice and maintain order, Western notions and institutions of law and justice were established, thus undermining and devaluing native institutions. United in its mission, a well-orchestrated bureaucracy wielded enormous power all over the urban and rural centers of population and penetrated the Indian social fabric.

The Indian elite, in awe of their new masters, began to identify with the new culture, and by identification with the values of their formal masters, viewed their own culture as deficient (Nandy 1983). Native institutions of learning, of healing, of justice and administration thus gave way to the imported variety; the British began discharging their “white man’s burden” in earnest. In 1835 the English language was introduced as a medium of instruction in place of the vernacular in Indian schools, and, for the first time, colleges using the English language were founded. The purpose was twofold: (1) to supply the British Raj with an army of clerks to supply the infrastructure of the bureaucracy and (2) to educate Indians in Western ideas, thus undermining the ancient culture they wished to replace. Heretofore, many Indians had studied English privately and had even traveled to England to establish their credentials as sophisticates among the alien rulers, but now an attempt to infiltrate the Indian culture began on a larger scale.

English manners and attitudes were emulated enthusiastically by the “upper crust” of Indian society. From within the ranks of this emerging class of Indian intellectuals versed in English manners and tastes arose not only a critique of Indian traditions, but also an attempt to confront inherent contradictions of a Raj, with all its complexities and faults.

Raja Ram Mohan Roy (1744/1833), from Bengal, was one of the first among the Indian scholars to lead the new movement for assimilation with Western practices. The British had initially established its seat of power in Calcutta, the major city of Bengal on the eastern shore, and hence had early on made its impact on this province. Ram Mohan Roy, after an exhaustive study of both the Western and Eastern traditions and religions, attempted to meld the best of each tradition. He championed the cause of education in English for Indians while urging the inclusion of Indians into the higher echelons of the British army. He was also a staunch advocate of freedom of the press, and a variety of other causes important to Indians such as attaining more welfare for the peasants, especially protection from exploiting landlords.

He attacked the injustices of the judicial system established by the British. Foremost among the reforms he advocated was the drive to abolish the tradition of Suttee (the British spelling of the Indian word *sati*), the practice of self-immolation by a widow on the funeral pyre of her husband. He met with intense opposition from Orthodox Hindus but was ultimately successful when in 1829 the British banned the infamous practice, most prominently practiced at the time in Bengal (Nandy 1980). Critical as he was of the Indian conditions and traditions, he ushered in an era of proud Indian voices that were not only seeking to reform the Indian society but also calling attention to the deficiencies and injustices of the Raj as well.

Other criticisms were directed toward the caste-ridden and divided Hindu society, which did not offer to all its members’ equal access to God, most glaringly to the Untouchables, who could not gain access to the Hindu temples. The Hindu leaders responded to these criticisms by forming social and religious units that assimilated the teachings of other religions, especially Christianity and Islam. Rejecting elaborate Hindu rituals dominated by the Brahminical order, societies were formed to integrate Hindu teachings and practices with those of the new religions. In 1828 Rammohan Roy helped establish in Calcutta a society called Brahmo Sabha, which took the form of an assembly of worshipers that rejected traditional Hindu temple worship The new movement had a substantial following among the Bengali elite over the next three decades and was later organized as the Bharmo Samaj, an organization devoted to social transformation.

A similar movement was organized in Bombay in 1849, the Prarthana Samaj, the name deriving from the gathering for prayers, akin to Sunday church services. The avowed purpose of these religious societies was to strive for social reforms, including advocating widow remarriages, assisting in the organization of night schools for adult education, and the establishment of orphanages. Those societies had as one of their goals the uplifting of the depressed classes throughout India. In all of the activities, the Indian societies were in competition with the Christian missionaries so as to be identified as the agents for social change throughout the country. A more fundamentalist thrust came from the Arya Samaj movement founded by Dayanand Sarswati (1824/1883), which sought to assert the primacy of the Vedas–the ancient Hindu scriptures–in religious discourse.

The Arya Samaj maintained that all the wisdom found in later tradition and in other world religions could be found in the Hindu scriptures. It was an attempt at raising the self-worth of the Indian people, wounded from the insults first from the Islamic conquest and the later humiliation at the hands of the Western Christian missionaries and rulers. Appealing to the nascent nationalism, which they hoped to enhance, the leaders of the movement preached the splendors of unadulterated Hinduism of the earliest Indo-Aryan tradition. Vedas, the earliest religious speculations of the ancient poets and philosophers that had been taken by the Hindu tradition to be divinely inspired, gave the movement an all-India appeal. Emphasizing monotheism and decrying image worship, the movement set out to purify the Hindu soul and society and pushed for social reforms. In harmony with other societies in India, the Arya Samaj rejected the restrictions of the caste system and championed and argued for choices for women both in matters of education and marriage (Parekh 1989).

The Theosophical movement, another socioreligious movement in India espousing freedom from ancient strictures and bringing together different religious ideologies and tenets in a socially open and inclusive manner, emphasizing reflection over ritual, made its imprint on the national scene in India in the late nineteenth century. Madame Blavatsky, one of its cofounders, came to India in 1879 and immediately became a part of the Hindu religious revival movements. Another person important in the religious revival in India, Annie Bessant, joined the Theosophical Society in 1889 and settled in India to participate in its religious and political rejuvenation. She was elected President of the Indian National Congress in 1917.

The efforts of the Indian leaders of the Theosophical movement led to many legal reforms, including legitimizing marriages of widows and legislating for an acceptable age of consent for marriage (Majumdar et al. 1980).

In the mid-nineteenth century, religion and social reform movements formed in India both in emulation of and in opposition to the values of the teachings of the British Empire which had colonized India since the eighteenth century. These movements were of value in enhancing the flagging self-regard of the Indians, so flagrantly disregarded by the British colonists. The various societies, in their attempts to rejuvenate and modify the ancient traditions, provided an infusion of worth to the Indians. Yet another contribution of these reform societies and movements was to till the soil for later political seeding-circa 1915—when Gandhi returned to India and began the march to Swaraj (Freedom) from the British Raj.

8.5.2 The New Rulers

The British colonists experienced little difficulty in establishing their control over the Indian states and principalities in a subcontinent that never had a central political administration and certainly had never had a cultural or religious center, diversity in all matters being the rule for India. A central feature of the British colonization was the impact they were able to make on the institutions indigenous to the many diverse subcultures in India. In various professional media—law, medicine, communications networks, and the transportation systems—attacks were made on Indian strategies and tactics, rooted in centuries of tradition, paving a way for the strategies and tactics of the British.

Thus the practice of law and the dispensation of justice underwent dramatic changes with the impositions of the British "masters." Never again did the country return to tradition-bound practices of settling disagreement and other legal matters through caste based adjudicating and interpreting bodies, rooted in caste-specific beliefs and practices. In fact, the law—British style—became a highly admired profession that was considered an important avenue to success by many segments of the Indian culture. It had been held out as an important avenue to success by many segments of the Indian culture, for example for the young Mohandas Gandhi (later lauded with the title *Mahatma*, meaning a great soul), important enough for him to journey to England and leave his mother, wife, and child. The Nehrus, Motilal and Jawaharlal, were highly respected members of the law profession in India; they practiced "British" law as luminaries in their subculture, thus endorsing the British ways. The Western medical profession was similarly respected and rapidly displaced traditional Indian medical practices.

Once again, the traditional could not withstand the attacks of the "modern" approaches. The traditional practices of the professions were vulnerable to the incursions of the new, "superior" ones and were easily replaced. Each of these movements marked a certain transition in Indian tradition, its ethos and its practices.

Another outcome of the sociological and ultimately psychological invasion of India was its impact on the lower classes. As we have noted, those of the upper classes in India who could take advantage of the "improvements" offered by the British did so while the poor and the illiterate were increasingly left behind, ultimately enlarging the division between the haves and the have-nots in the country. The underclass had no place in the new society that was forming; indeed for many decades their lot was worsened by the "improvements" throughout the land.

Mahatma Gandhi returned to India from South Africa in 1915 and within a span of a few years became not only a political but also a spiritual leader of India. His major goal was independence but, equally important to him, was the essential restoration of dignity to all Indians, women alongside men, untouchables alongside the Brahmin, the rural masses alongside the urban elite.

He most famously championed the cause of the untouchables, and although it cannot be said that untouchability has been eradicated, the clock cannot be turned back on the status of the untouchables either. Unfortunately, because of the strong protection afforded by the constitution and the vagaries of electoral politics, although the lot of the untouchables and other backward classes has improved greatly and they feel empowered, the status of the work itself has not changed, and certain tasks remain in the exclusive domain of these former backward classes.

Another Gandhian idea is that of "trusteeship," aimed at mediating between the capitalist and laboring classes. Gandhi wanted the owners of the means of production to become trustees of their estates rather than owners. He tried to skirt the ideas of socialism at the same time, appealing to the conscience of the entrepreneurial classes to assume responsibility for the wellbeing of their workforces. The idea was aimed at the economic order, such that traditional values of caste and kinship organization could be brought to bear upon modern political and economic transactions. This was one of a few ideas that Gandhi tried to advance that went nowhere.

But the spirit with which he imbued the independence movement, the inspiration both from within the tradition and from ideas of Western (particularly Christian) liberation with which he tried to inform the new independent state of India, and the western looking leadership that he had mentored, saw to it that the first task of the state was to address the inequalities that bred vulnerabilities among the people of India.

8.5.3 *After Independence*

Not too long after independence in 1947, famine was gone as the result of a democratic distribution system, and in another short while smallpox was eradicated as well, thus eliminating two great scourges to which the Indian population had been vulnerable. With a new constitution, India launched upon a drive to level the playing fields for scheduled tribes and scheduled castes, particularly the untouchables.

New irrigation projects focused on the need for feeding millions of India's poor, and an invigorated health care delivery system with emphases on primary care was a sea change for the rural masses. Over a few decades significant reductions in infant and maternal morbidity and mortality were achieved. Female literacy and employment led to a reduction in birth rate, but a rapidly expanding young population put new demands on the system.

In the last three decades the economy has been growing rapidly, and with it, more resources have become available to focus on immunization, disease prevention, and availability of safe drinking water. Multiple forces have incrementally changed the status of the rural masses in India. In the nineties India opened its economy, accelerated its growth, and globalization as well as the age of information have lifted significant numbers of rural and urban masses from the onerous burden of poverty. The Indian pharmaceutical industry has blossomed, becoming a supplier of all manner of drugs not only to Indians but also to other third world countries. India is opening up as a place for medical tourism with its state of the art modern hospital available to those who can afford it.

Access to health care still remains a major ethical problem. Maldistribution of health care professionals is even worse than in the Western world, and is a far cry from providing for the needs of millions of urban, rural, and remote populations. On top of any other vulnerability acquired as a result of India's stratified social structure, the burden of poverty is simply backbreaking. Deprivation of every known human need is the norm still for several million Indians. Even so, the taste of democracy is ever so sweet and the population at large enthusiastically embraces adult franchise, and savors the autonomy the new system has emboldened.

8.6 Bioethical Challenges for the Twenty-First Century

As is clear from even a cursory examination of the post-independence changes, most changes have been inspired by Western liberal values, particularly those imbedded in democratic governance with universal franchisement. The challenges come from the resultant conflict between these rapidly changing aspirations of the people and customs and practices rooted in traditional value systems.

First let us examine the issue of autonomy, not only in health care decisions but also in life in general. As a hierarchical society, India is also a patronage society. The young are expected to defer to the old; in return it is incumbent on the elderly to protect the young. As the upper castes dominate and expect services from the lower ones, they must provide, however meagerly, for those who serve them. Women must obey their fathers and husbands, but in turn their well-being must be a priority for the family. This is not as idyllic as it may seem because not only does practice violate these principles day in and day out, but their dependent and servile status deprives the young, women, and the lower castes of the autonomy to make decisions for themselves.

Young people, who are encouraged by education, information, and liberal values to make their own career as well as marital decisions, often confront the older, more conservative wishes, of their elders. As for health care, it has not quite yet dawned on the larger Indian population that these decisions should be made by individuals autonomously. As anthropologist McKim Marriott (1990) has pointed out, Hindu persons are not understood to be individuals but rather "dividuals." The skin is not the boundary of their self or personhood and interpersonal transactions are governed by dynamics of high and low, hot and cold, heavy and light, etc., thus making the Hindu person permeable. Thus it is assumed that elders in the case of the young (as one would treat preadolescent children in the West), and men in case of women, will make health care decisions. Formal consent is not even an issue, and often not only the young, but adult females and lower caste persons, are treated without consent, and sometimes under false pretenses and even duress. The idea of informed consent is practically unknown. Indian tradition values the ideal of a support system far more than it does Western notions of individual autonomy.

It is routine for a physician to withhold an adverse prognosis from the patient, especially when this is the wish of the family. The Western idea of the person's body being his/her fortress or his/her home is alien in India (homes too are open and the traditional word for a guest is 'atithi,' meaning 'the date of whose arrival is unknown'). To force an injectable drug into a person is more common than the exception. Hospitalization of the mentally ill often takes place against their will, without resort to a court of law. Unregulated sectors of pharmaceutical research are allegedly beset with the same problems of consent.

Poverty also makes informed consent and the idea of justice and beneficence problematic. The poor are forced to sell their organs, and sometimes their children, into bondage, servitude, or prostitution. A few instances of abuse of domestic help (called and treated as servants in India) have been reported even among Indian diplomats serving abroad, not to speak of plentiful examples in India.

Modern laws ushered in by India's constitution do not embody the values of all people and legal redresses are not so easily available. It is said, not in jest, that a civil case filed against someone may get resolved in the lifetimes of one's grandchildren. A major source of corrective influence is thus essentially absent. Older institutions, like caste-based jury-like panels do exist, and do address some of these problems in the closed context of a particular caste or village, but these are fragmenting in a rapidly urbanizing India. In some instances these bodies have revolted against the rapid advance of democratic values and taken matters into their own hands, particularly when it is a matter of marital decisions of the young that go against established norms of a caste society.

As noted above, modern technology has also thrown up a particularly bothersome challenge, that of sex-selection. There is an overwhelming preference for male children, with females seen as being burdens on their parents, children who grow up only to make someone else's home and that too at a major parental cost. Combined with prenatal diagnosis of gender, the wish to have male children has made female fetal abortions a matter of national alarm. Again, in some states of India laws have been enacted banning the practice of prenatal gender diagnosis, but since abortion itself is legal, the laws against it have had little preventive effectiveness.

In addition, medical technology has given rise to an unregulated marketplace for surrogate motherhood, with it its legal and emotional complications, both among some Indians and many outsiders troubled by childlessness.

8.7 Conclusion

The values espoused by the UNESCO declaration are essentially Western. Although India as a political entity has embraced Western liberal values of adult and universal franchise and democratic governance, it can hardly be said that the values inherent in these social and political institutions have anywhere near completely permeated into the Hindu society and psyche. Clearly, there are winners and losers in electoral politics and the tendency is to sway elections by using traditional connections, if not outright through corruption. Laws are far from being able to provide a structure that can contain conflict, and there are not adequate resources yet to redress man-made inequities and vulnerabilities. In spite of significant progress in the health delivery system, the benefits of the modern era are yet to reach millions.

In conclusion, transplantation of values is not an easy process. Tradition conflicts with modernity and gives way only with difficulty. There are more apparent losers in a traditional society than there are winners, but the last 60 years have seen only a further determination on the part of India's leaders and some of its people to strengthen these new values.

References

Sushrutasamhita (Text and Gujarat translation) Trans. Shastri Govindaji K. Ahmedabad. SSV Karyalaya 1973.

Hutchins, F. 1967. *The illusion of permanence, British imperialism in India*. Princeton: Princeton University Press.

Majumdar, R, H. Raychaudhari, and K. Datta. 1980. *The advanced history of India*. London: McMillan.

Marriott, M. 1990. Constructing an Indian Ethnosociology. In: *India through Hindu categories*, ed. M. Marriott. New Delhi: Sage.

Nandy, A. 1980. *At the edge of psychology: Essays in politics and culture*. Delhi: Oxford University Press.

Nandy, A. 1983. *The intimate enemy*. Delhi: Oxford University Press.

Nehru, J. 1981. *The discovery of India*. New Delhi: Jawaharlal Nehru Memorial Fund.

Parekh, B. 1989. *Colonialism, tradition and reform*. New Delhi: Sage.

Radhakrishnan, S, and C. Moore. 1973. *A source book in Indian philosophy*. Princeton: Princeton University Press.

Caraksamhita (Text and English translation) Trans Priyavrat Sharma. 2 vols. Varanasi: Chauakhambha Orientalia, 181–83.

Singh, JP. 2010. Socio-cultural aspects of high masculinity ratio in India. *Journal of Asian and African Studies* 45:628–644.

Spear, P. 1978. A History of India. 2 vol. Harmondsworth: Penguin Books.

Chapter 9
Human Vulnerability in Islam

Mustafa Abu Sway

9.1 Introduction

> The Prophet frowned and turned away,
> Because there came to him the blind man, [interrupting].
> But what would make you perceive, [O Muhammad],
> that perhaps he might be purified,
> Or be reminded and the remembrance would benefit him?
> (Qur'an, 80:1–4)

In this Quranic verse we read of a vulnerable person, a blind man whose name is Abdullah Ibn Um Maktoum (d. 637 CE/15 AH), approaching Prophet Muhammad (Peace be upon him) while he is attending the affluent and powerful Meccan leaders hoping that they will embrace Islam, but the Prophet frowns and ignores the blind man. Muslim scholars cite this verse in polemical discourses as a proof for Muhammad's prophethood and the Qur'an being the revealed word of God, for if the Qur'an were the Prophet's own creation, he would not have included such an event.

Indeed, this Divine intervention on behalf of those who are vulnerable would not go unnoticed or without action. One can detect a clear affirmative action that the Prophet took to accommodate Ibn Um Maktum.

According to the great jurist Ahmad Ibn Hanbal (d. 855 CE/241 AH) in his *Musnad*, the Prophet put Ibn Um Maktoum in charge of leading the prayer in Medina twice. Abu Dawud (d. 888 CE/275 AH) listed this event in his *Sunan* under the subtitle "The Blind Imam."[1] Ibn Um Maktoum also used to make the call for prayer in the presence of the Prophet.[2]

Al-Tabari (d. 922 CE/310 AH) says in *Jami' Al-Bayan fi Tafsir Al-Qur'an*, in his commentary on chapter 80 (Surat 'Abasa) of the Qur'an, that as a result of this

[1] Abu Dawud. Sunan. hadith #595.

[2] Muslim, Sahih. hadith #381.

M. A. Sway (✉)
Integral Chair for the Study of Imam Al-Ghazali's Work, Holy Al-Aqsa Mosque and Al-Quds University, Jerusalem, Palestine
e-mail: abusway1@gmail.com

J. Tham et al. (eds.), *Religious Perspectives on Human Vulnerability in Bioethics*, Advancing Global Bioethics 2, DOI 10.1007/978-94-017-8736-9_9,

revelation the Prophet was very kind to Ibn Um Maktum thereafter. The most important aspect related to our topic here, however, is that this revealed chapter supports the claim that those who are vulnerable should be given at least equal rights, if not more. They should be respected and provided with the same opportunities, similar to healthy members of the society whenever possible. All human beings need to be respected and their uniqueness recognized, for suffering from impairment does not lessen one's humanity.

It is obvious that vulnerability forms a false sense of otherness which could lead to discrimination. Otherness could be defined in terms of gender, culture, ethnicity, and socio-economic and political background, to name a few. Examples include women in general, widows and divorced women, those who suffer from calamities, those in debt, and some Muslim scholars include non-Muslim minorities as well. In this paper, otherness is defined in terms of vulnerability where human beings suffer from a temporary or permanent condition of weakness, sickness or physical disability, where they are susceptible to ill-treatment, be they physical, emotional, or both.

While I do advocate a theology of soft-otherness in relation to the theological other (Jews and Christians in particular), where health care is concerned it is not easy to adequately address the ontological status of the human being in relation to her dignity. Otherness, when it paves the way for discrimination in medical care, should be deconstructed. The common denominator is humanity, and all people should have access to the same health care.

9.2 Vulnerability and Weakness: Our Human Condition

It is only natural that human life goes through different phases of physical well-being, from the weakness of infancy and childhood to the strength of youth, then again to the state of being old and feeble:

> Allah is the one who created you from weakness, then made after weakness strength, then made after strength weakness and white hair. He creates what He wills, and He is the Knowing, the Competent. (Qur'an, 30:54)

But even the strength of youth is relative to other human beings, if human conditions are seen against God's Omnipotence. Humanity, in its relationship with God, is in a state of deficiency and powerlessness, in need of material and spiritual nourishment: "O mankind, you are those in need of Allah, while Allah is the Free of need, the Praiseworthy." (Qur'an, 35:15)

Human weakness is used in the Qur'an and the Prophetic traditions to justify relaxing the law including fasting, to accommodate those who cannot fast either temporarily or permanently:

> The month of Ramadhan [is that] in which was revealed the Qur'an, a guidance for the people and clear proofs of guidance and criterion. So whoever sights [the new moon of] the month, let him fast it; and whoever is ill or on a journey—then an equal number of other days. Allah intends for you ease and does not intend for you hardship and [wants] for you to complete the period and to glorify Allah for that [to] which He has guided you; and perhaps you will be grateful. (Qur'an, 2:185)

Those who cannot or should not fast include elderly people whose health could suffer tremendously if they fast. Islamic law also allows room for certain acts during the Hajj not to be performed if one has certain medical conditions. Those who suffer from permanent adverse medical conditions have recourse to penance, required only from those who are able financially, which in turn helps poor people.

There will always be a measure of weakness and vulnerability in everyone. That is why sickness and death are not restricted to the old. In the shadow of death, when an end to this life is simply remembered, one recalls his own mortality and vulnerability. Therefore, the less vulnerable is called upon to serve those who are more vulnerable.

9.3 Children and Women: Rights and Needs to be Protected

The Prophet himself suffered as an orphaned child, and got a taste of what it means to be vulnerable. He stressed taking care of the material and emotional needs of orphans, to compensate them for their loss, he said: "Bring the orphan close to you, be kind to him, touch his head and feed him from your own food. This will soften your heart and fulfill your needs."[3]

According to the Qur'an, taking care of the orphans and other vulnerable persons was part of the original covenant between God and the Children of Israel:

> And [recall] when We took the covenant from the Children of Israel, [enjoining upon them], "Do not worship [anyone or anything] except Allah; and to parents do good and to relatives, *orphans*, and the *needy*. And speak to people good [words] and establish prayer and give alms." Then you turned away, except a few of you, and you were refusing. (Qur'an, 2:83)

The fact that the history of revelation, including the Torah and the Qur'an, addressed the needs of those who are vulnerable, points to the universality of their rights and the need to protect them.

The Prophet also stressed the rights of children and the elderly, with the underlying message that these are two general categories of vulnerable people, stating that "He is not one of us who is not merciful to our young and does not respect our elderly."[4] The reward for taking care of orphans is being in paradise in the immediate company of the Prophet.[5]

The rights of one are the duties of others, and if one cannot fulfill these rights, it becomes the duty of a larger circle of people until it involves the whole society. But these rights and duties are decided through revelation (i.e., the Qur'an) which, in turn, gives legitimacy to the Sunnah of the Prophet. The latter is reflected through compendia of Prophetic traditions, the most authentic of which are *Sahih Al-Bukhari* and *Sahih Muslim*.

---

[3] Al-Bayhaqi. Sunan. 4/60, #6887.

[4] Al-Hakim. Al-Mustadrak, 1/62; Al-Hamidi, Musnad, vol 2, hadith #586.

[5] Al-Bukhari. Sahih. hadith #5659.

Therefore, all commandments, recommendations and prohibitions have to be rooted in the Qur'an or the Sunnah, in order form part of the Islamic worldview. Muslim practices should not be seen automatically as a reflection of what Islam says, for it could be exactly the opposite. In the discussions that ensued after the Rome conference, there was a reference to female circumcision among the Islamic populations. According to the World Health Organization, Female Genital Mutilation (FGM) "is practiced by followers of different religions—including Muslims, Christians (Catholics, Protestants and Copts), and Animists—as well as by non-believers in the countries concerned" and the WHO statement on the origin of FGM declares that "It is not known when or where the tradition of female genital mutilation originated."[6] Nevertheless, WHO statistics show that the overwhelming majority of women suffering from FGM are in Africa.

The Prophet did emphasize taking care of women in his last sermon in Mecca. He wanted to make sure that the Muslim community would continue his Sunna, which liberated women from the ignorant pre-Islamic practices.

Muslim women owned property, led other women in prayers, competed with men in the spiritual realm, and enjoyed the fact that house work became genderless because the Prophet himself did household chores. Most importantly she was liberated from the burden of responsibility for what happened in the garden, because the Qur'anic narrative uses a suffixed dual pronoun when referring to Adam and Eve being both warned, both tempted, both ate and both expelled from the garden. None of them was *first* at any stage in the story of the Qur'an.

9.4 Extending Mercy to the Elderly and the Sick

The very last stage of human life may entail many health problems, physical and mental, leading ultimately to complete loss of memory:

> And Allah created you; then He will take you in death. And among you is he who is reversed to the most decrepit [old] age so that he will not know, after [having had] knowledge, a thing. Indeed, Allah is Knowing and Competent. (Qur'an, 16:70)

Prophet Muhammad himself used to seek refuge in God not wanting to reach that decrepit old age.[7] One may think of all the infirmities addressed by geriatrics as a way of interpreting decrepit old age, yet Al-San'ani (d. 1850 CE/1182 AH) interpreted decrepit old age in *Subul Al-Salam*as "dementia."

Amongst those who usually get old are parents. The value of respecting parents and those in the same category, including grandparents and, by extension, other elderly persons, is universal.

[6] Female Genital Mutilation. World Health Organization. http://www.who.int/gender/other_health/Studentsmanual.pdf, Accessed 4 May 2012.

[7] Al-Bukhari. Sahih. hadith #6013.

The Qur'an mentions being good to parents next to stressing the oneness of God, for any act that violates the rights of parents, especially when they get very old, cannot be reconciled with having a healthy relationship with God:

> And your Lord has decreed that you not worship except Him, and to parents, good treatment. Whether one or both of them reach old age [while] with you, say not to them [so much as], "*uff*," and do not repel them but speak to them a noble word. And act humbly to them in mercy, and say, 'My Lord, have mercy on them, since they cared for me when I was small (Qur'an, 17:23–24)

This verse reflects a zero-tolerance theology concerning the mistreatment of parents, especially when they become elderly. This theology is extended to all the elderly, family or not. The mistreatment of parents is considered a grave sin in all Islamic schools of jurisprudence. While good treatment is mandatory, children cannot pay parents back what they have done to them, especially the mother. Abdullah Ibn 'Umar saw a companion of the Prophet performing pilgrimage in Mecca while he was carrying his mother on his back. He asked Abdullah Ibn 'Umar if he paid her back her rights with his action. He answered him saying that his deed was excellent but it does not pay the equivalent of one of her contractions [while giving birth to him], and that God would reward him abundantly for the small work that he has done.

> A man came to the Prophet and asked him: Who is entitled to my best companionship? He replied: Your mother! He [again] asked: Then who? He replied: Your mother! He asked Then who? He replied [for the third time]: "Your mother!" He asked: Then who? He replied: "Your father!"[8]

Some scholars interpret this tradition as saying that the first answer is in appreciation of the mother's pregnancy, the second answer is because of the pain of labor associated with delivering the child, and the third is because of breastfeeding.

Respecting elderly in general is highly regarded in the Prophetic traditions. According to Al-Tirmidhi, the Prophet (peace be upon him) said: "No young man that honors an old man, out of respect for his old age, except that God will send him someone to honor him when he reaches old age."[9]

Al-Hakim narrated in his *Mustadrak*that the Messenger of God used to visit the feeble and sick Muslims. In addition, Al-Bukhari narrated in his *Sahih*that the Prophet also visited the dying son of a Jewish neighbor.[10]

The companions of the Prophet followed his Sunnah in taking care of those who were vulnerable. Ibn Al-Jawzi narrated in the *History of Umar Ibn Al-Khattab* that Talha, a companion of the Prophet, saw 'Umar going at night to two different houses. When Talha inquired [in the morning] he found a blind old woman. He asked her: "What is the story of this man who comes to you?" She said: "He has been serving me for quite some time; he brings me what I need, and removes the trash [from the house]…"

---

[8] Narrated by Al-Bukhari (#5626) and Muslim (#2548).

[9] Al-Tirmidhi. Sunan. hadith #2022.

[10] Al-Bukhari. hadith #1356.

Abu Bakr Al-Siddiq used to compete with ‘Umar in serving the same old lady. Suffice it to know that we talk about the first and the second caliph respectively. This is the ethos that prevailed in the formative years of Islamic history.

In the history of the Umayyad dynasty, Caliph ‘Umar Ibn ‘Abd Al-Aziz sent a letter to ‘Adiyy Ibn Arta’a, the governor of Al-Basra in Iraq, telling him to look after the Jews and the Christians “who became old, weak and cannot make a living” and to provide them with regular income from the treasury of the state.[11] He also asked his deputies to send him the names of all the blind, handicapped, paralyzed or he who is permanently ill to the degree he cannot get up to pray. He sent a guide for every blind person and one employee to take care of two special needs people.[12]

The Umayyad Caliph Al-Walid Ibn ‘Abd Al-Malik went one step further and he established an institution to take care of those with special needs, employing physicians and servants, and there was regular payment for everyone including those with special needs.[13]

That the relationship to others based on extending mercy to all is captured by the Andalusian Maliki scholar Ibn Battal (d. 1057 CE/449 AH) who commented on a Prophetic tradition that promotes being merciful[14] and which was narrated by Al-Bukhari, he said: “It encourages extending mercy to all creatures, the believer and the non-believer, and domestic and wild animals…”[15]

9.5 Medical Care and Bioethical Issues

Medical care in Islam could be divided primarily into two categories: First, the imperative in favor of treatment when medicine decrees that *not* seeking treatment is detrimental to health and that the treatment would usually work. Second, it is permissible to *not* seek medical treatment that is not detrimental to health, or that treatment is not capable of improving the health of the sick person. That it is permissible means one has the option of not seeking medical treatment.

Al-Tirmidhi narrated that when the Bedouins asked the Prophet whether they should seek medical care, he replied: “Yes, O servants of God! Do seek medical treatment, for God created a medicine for every disease, except old age.”[16] As for those who find themselves in a very difficult situation, the Prophet advised them not to wish for death, he said: “None of you should wish for death, nor supplicate for it before it reaches him. Verily, once one of you dies, his [good] deeds come to an end, and the age of the believer only increases the good.”[17] No degree of pain

[11] Al-Qasim Ibn Salam. (1989) Al-Amwal. In: (ed) Muhammad `Amarah. Beirut, Dar Al-Shuruq, 121.

[12] Ibn Al-Jawzi. Sirat `Umar Ibn Abdulaziz, p. 130.

[13] Ibn Kathir. Al-Bidaya wal-Nihaya, p. 186.

[14] Al-Bukhar.i hadith #5583.

[15] Al-Mubarakfuri. Tuhfat Al-Ahwadhi bi-Sharh Jami` Al-Tirmidhi. hadith #1922.

[16] Al-Tirmidhi. hadith #2038.

[17] Muslim. hadith #4843.

should lead the elderly or the very sick person to seek active euthanasia or assisted suicide: "… And do not kill yourselves [or one another]. Indeed, Allah is to you ever Merciful." (Qur'an, 4:29)

Not only the patients should not seek to end their life, health care providers should not coerce them to do so for they are not more merciful to them than their Creator. Life and death are matters that belong to God: "And it is He who gives life and causes death, and His is the alternation of the night and the day. Then will you not reason?" (Qur'an, 23:80)

Life and death issues, coupled with the advancement in medical technology, pave the way for experimentation, active euthanasia and genetic engineering, to name few possibilities where the rights of the vulnerable could be violated. Reasons such as cost of medical care, and social security/pension should never be allowed to play a role in putting pressure to legalize euthanasia, as an example, because the elderly or the sick are considered a burden on the state resources.

From an Islamic point of view, IVF is allowed to help married couples who are infertile, but those outside of such relationships are not eligible for such help The practice of surrogate motherhood derived from IVF is prohibited in Islam. When this is done for commercial purposes, people use their bodies and are only trustees. Islamic Shari'ah prohibition of this practice stems from the aims to protect the progeny. Another limitation comes from the religious ruling that extra embryos should be allowed to "die." It should be known that these embryos exist in a pre-fetal and, therefore, a pre-ensoulment stage. They are not considered human beings and they do not have legal rights as fetuses. There is no central authority amongst Muslims and it is almost inevitable that Muslim scholars would differ. Stem cell research has limitations that include prohibition of egg or sperm donations where extra embryos obtained from infertility "therapy" find their way to research, sometimes without the consent of the "donors." Poverty might create conditions of vulnerability where people are pushed to "donate" sperm or eggs, and a combination of riches and infertility might lead others to seek those "donors."

When it becomes apparent that medicine cannot help, one needs to remember that there is a Divine Will regarding these issues:

> To Allah belongs the dominion of the heavens and the earth; He creates what he wills. He gives to whom He wills female [children], and He gives to whom He wills males. Or He makes them [both] males and females, and He renders whom He wills barren. Indeed, He is Knowing and Competent. (Qur'an, 42:49–50)

Cloning full human beings will always be problematic from an Islamic perspective, even when it is not done for "spare parts"! The nature of the relationship with the source of the 46 chromosomes is not well defined. More accurately, there is no category for a cloned person in Islamic Shari'ah. He or she is not a son or a daughter in relationship with the source! He or she is simply a clone, which would prove to be an alienating factor leading to psychologically distressing relationships, probably worse than the rejection that the fictional Dr. Frankenstein had towards his "monster." What is monstrous in cloning a human being is the act, not the shape of the final "product"!

The vulnerable persons also includes those who are susceptible to worldviews that convince them to change their nature as being fine. There is a verse in the Qur'an that addresses the prohibition to change the human body which I understand to include transsexual operations, the procedure to change one's appearance from male to "female" or from female to "male" based on one's "feelings" of being the other sex: "And I (i.e., Satan) will mislead them, and I will arouse in them [sinful] desires… and I will command them so they will change the creation of Allah…" (Qur'an, 4:119)

9.6 Concluding Remarks

Human beings gradually pass, in most cases, from celebrating their youth and health to a condition that lacks both. It is inevitable that many become vulnerable and need special care, but above all, they need respect and dignity. The cornerstone of the relationship with fellow human beings in Islam is based on justice and the recognition of their dignity. The Qur'an shows that God honored humanity at large: "And We have certainly dignified the Children of Adam…" (Qur'an, 17:70) According to Al-Razi (d. 1209 CE/606 AH) in *Mafatih Al-Ghayb*, a major aspect of this dignity is that human beings are endowed with the intellect that enables them to know God.

Vulnerability could be taken to a different level. According to the World Bank, in 2008 there were 1.29 billion people in the developing world who lived on less than $ 1.25 a day. There is a direct correlation between poverty and vulnerability. Lack of food, health services and education is appalling. The discrepancy between the north and the south should be bridged and a redistribution of wealth is imperative.

While it is possible to redress the gap between the north and the south, and to take care of the vulnerable amongst us, by appealing to our shared humanity, it is important to remember the role of all messengers and to restore the Prophetic model of mercy (Qur'an, 21:107), for that is exactly what is needed today.

Chapter 10
Reflections on Human Vulnerability and the Rabbinic Perspective on Medical Ethics

Yechiel Michael Barilan

10.1 Introduction

Like care and responsibility, the value of human vulnerability has the intuitive appeal of an old truth even though the concept is relatively new and its meanings far from being elucidated.

Vulnerability is also a derivative concept, as it receives its calling from human dignity and human solidarity. Many vulnerable things exist out there in the world, yet we do not care too much about them. The vulnerability of things that have a special moral status (e.g. dignity) and a special moral relationship to us (e.g. solidarity) is already a descriptive adjective soaked with normativity.

Vulnerability is also beset by an internal contradiction, because many callings need to be considered and the protection of vulnerability entails some risks and sacrifices as well. If we underscore our vulnerability too much, we will be less inclined to undertake altruistic humanitarian missions or innovative enterprises. Notable examples are blood transfusions, medical experimentations and vaccinations, which have raised much hostility due to the risk involved for people, especially children. Although the ethical standards then, at least in the eyes of the professionals, were different, i.e., less stringent than today, it is evident overall that caregivers and patients alike are not so vulnerable as to exclude any risk-taking that bears promises for saving life and for their own betterment. Attunement to human vulnerability usually entails reliance on human strengths, such as some physical resilience and will power.

Not disclosing "truth" or hiding "bad news" from critical patients was, and in many places still is, a typical practice which today we behold as an excessive worry about vulnerability. The Jewish Code of Law *Shulhan Arukh* (YOREH DE'A 337) says that it is prohibited to communicate to a patient the death of a person close to him, for fear of "terrifying his mind." The probable moral presumption is that people want to be treated this way when acutely ill. The naturalist presumption in

Y. M. Barilan (✉)
Sackler School of Medicine, Tel Aviv University, Tel Aviv, Israel
e-mail: ymbarilan@gmail.com

J. Tham et al. (eds.), *Religious Perspectives on Human Vulnerability in Bioethics*, Advancing Global Bioethics 2, DOI 10.1007/978-94-017-8736-9_10,

Jewish law is that critically ill patients are highly sensitive to emotional distress (Barilan 2014, p. 32, 74). This example highlights vulnerability as both biological and cultural. In Jewish law, the biological vulnerability of seriously ill people is demarcated, protected and, perhaps, also enhanced, by a normative cultural conceptualization of the value of life.

No less challenging is the observation that social practices and institutions have been the chief protectors of the vulnerable, yet they have also both created vulnerabilities and abused them. A notable example is the vulnerability of homosexuals and transgendered people to which the Abrahamic religions have contributed so much. This point looms large over the dimensions of vulnerability this book focuses on—gender, procreation and old age. The predicament of gay and transgendered people has not been resolved with the abolition of persecution and with the *de facto* negative liberty granted them by the secular state system. Rather, without the capacity to marry and to avail themselves of infertility treatments and adoption services, they are doomed to marginalization and to the vulnerabilities of lonely ageing, which are strongly associated with significant levels of morbidity, mortality and unhappiness relative to people with families (Bok 2010, Ch 10).

Although this paper does not address the problems mentioned above directly, it is crucial that we set them clearly in the forefront of our moral consciousness. The vulnerabilities our moral traditions already acknowledge are more protected somehow, even if not satisfactorily, than the vulnerabilities still in the shadow. The genuine moral challenge posed by the value of vulnerability is the psychological, moral and political resistance to acknowledge vulnerabilities for which we are responsible. Recognition is the key word in the value of vulnerability.

Religion has always been responsible for another kind of vulnerability, that is, dissent. "Heresy" or "disrespect" for enshrined values has justified disregard, silence, discrimination, and even direct violence. We have to bear in mind that every attempt to point out an *unrecognized* vulnerability and call for its defense is likely to be labeled as dissent and targeted for oppression. Hence, the key respect for the value of vulnerability is freedom of conscience, expression of critical reasoning, and openness to change.

Nonetheless, reflection on different traditions and doctrines helps us see bedrocks of moral common denominators, even under seemingly unresolvable and heated conflicts such as between beneficence and autonomy or the moral status of the unborn. Digging deep and critically might help contemporary multicultural society to construct effective and moral structures in defense of human values and vulnerabilities.

In this paper I will first explore philosophically the notion of human vulnerability and then examine a few contemporary rabbinic rulings that in my opinion reflect a distinct conceptualization of human vulnerability regarding the sick, women and early human life.

10.2 Vulnerability as a Philosophical Concept

The ethos of human dignity, which cherishes the life and flourishing of every human individual equally, for all of humanity and throughout the entire life-cycle, is evidently in conflict with the facts that all humans are constantly undergoing an inevitable and only partially predictable progress towards physical annihilation, and that humans vary from each other in their particular vulnerabilities. All humans are susceptible to all sorts of harms; no human is immune to basic harms such as illness or aggression. No human or ethnic group is significantly stronger or more resilient. Some humans are vulnerable to higher risk than others. But all humans are at risk of becoming victims of violence, and most humans have the power to bring about considerable harm to virtually every other human being within reach (Kavka 1983).

There is a very strong interdependence among humanity's basic interests and its vulnerabilities. For example, a tooth ache might deteriorate into lethal infection; the pain might mire the clarity of personal judgments (think of a prime minister taking fateful decisions while suffering from such pain); the medication given might have rare but devastating side-effects. Many of these harms are irreversible.

Additionally, people's capacity to predict and estimate harm is limited, even regarding themselves. Irreversibility precludes experimentation and replacement. It is even more difficult to predict and to estimate mental harm and its reverberating impacts on physical well-being and the social roles of the person. Not only do ordinary people, but even experienced rehabilitation doctors, underestimate the reported life-satisfaction of their severely disabled patients (Bach and Tilton 1994). Even when sick and disabled people report that they are unhappy, the causal factors are attributable to non-disability related hardships of life, which are often aggravated by difficulties at finding respectful accommodation on behalf of society, not to the disability itself (Albrecht and Devlieger 1999). Physicians tend to grossly overestimate the prognoses of their terminal patients, and to be even more optimistic when they inform these patients about their unrealistic prognosis (Lamont and Christakis 2001). The likely outcome is an excessive recourse to futile and often painful care. One of the most vexing experiences of care-giving is the inescapable duty to choose for others in matters of life and death.

Especially susceptible to such harm are self-esteem, rational judgment and the senses of meaning and willing. Overall, it seems that coercion of the person to act against his or her choice, especially when accompanied by a condescending attitude, is one of the gravest psychological harms possible. For instance, sexuality is especially vulnerable, physically as well as psychologically, in even short and ostensibly minor sexual abuse. Victims of sexual assaults describe these few minutes as "death" and "annihilation" that are irreducible to any other painful experience (Baker 1997, pp. 604–605).

Some mental faculties are quite beyond scrutiny. If we care for them, we must grant them extra-protection and we must do so through trust in one's own first person perspective. These are sincerity, trust, understanding, free will, good will, and sexuality related choices. Sometimes we have to wait until one has a voice to empower the person to self-reflect properly and speak up, or to sincerely assume the voice of those who cannot have one. Because the absence of remediable dependence on others in matters of personal life and self-care is an aspect of freedom and dignity, this interdependence also enhances vulnerability.

Because of the interdependency of vulnerabilities, the vulnerability of dependence on others' voices correlates highly with other vulnerabilities (e.g. mental retardation is associated with high risk of morbidity and mortality).

Our abstract mental faculties are expressible in bodily metaphoric idioms (Lakoff and Johnson 1999) and are vulnerable to prejudice and bias that are related to bodily conditions and perceptions. Pain and pleasure, even minor satisfactions and frustrations influence one's generosity in a given moment (Isen and Levin 1972). Minor alterations in the external framing of a situation have a considerable impact on people's moral actions, despite their readiness to save life (Darley and Batson 1973). Fair judgment is obfuscated by physical features such as beauty, height and even ethnic appearance. Category confusion of the moral and non-moral normative (e.g. social conventions, prudence), the aesthetic and the pleasurable are intrinsic features of the embodiment of the self (Haidt 2001). For example, people tend to identify cleanliness and whiteness with values such as "goodness" and "purity."

From these observations we may learn that the normative dimensions of human vulnerability are evasive, difficult to grasp and verbalize. This might be true regarding other moral values, but the challenge in vulnerability seems to be the paradoxical situation in which enhanced efforts to prevent vulnerabilities and protect the vulnerable also tend to enhance vulnerability, create new vulnerabilities and even, despite the best of intentions, harm and humiliate the vulnerable.

If we try to make the life of plants and animals better, we study them scientifically and control their conditions of life accordingly. But attempts to scrutinize the person objectively, or to decide for his or her own good, tend to result in failure and resentment since a person cannot be examined as an object like a plant or an animal. Even effective surveillance is not in line with respect for dignity. Attention, deliberation and persuasion are more respectful of the dignity of the persons involved, if even only partially effective, as sincere (loyal to self) and pragmatic (open to learning and virtuous transformations) agents. But these processes are doubly vulnerable, because they are sustained by the vulnerable cooperation of vulnerable persons, and by means of the elusive and vulnerable attitude of sincerity. As long as we cherish these vulnerabilities as part of our nature, and as long as we wish to cope with and not eliminate them, vulnerability is a special moral value.

Trust and good will are two of those very partially discernible, and hardly manipulable, mental states of mind on which we all depend. If my neighbor or the person riding the bus next to me either do not trust my own good willing or do not care about anything, even self-interest, there is little that can stop him from pulling out a gun and shooting me in the face. Our natural vulnerabilities draw humans to each other, thus exposing them to interpersonal hostilities, the chief bulwarks against which are goods that nobody can manufacture or purchase: trust and care.

In a broad moral paradigm that celebrates human life and values, the combination of vulnerability, interdependence of vulnerabilities, the epistemological shortcomings regarding vulnerability, and the ensuing vulnerabilities of sincere cooperation and interpersonal dependence encapsulate together the special duty to protect the embodied human person, especially regarding mental faculties that bear on free rational choice. We have to respect one's own voice, especially regarding identity and conscience, in order to protect one's embodied person properly. This recursive circularity enhances vulnerability to abuse, self-abuse and confusion. It renders sole reliance on the moral value of vulnerability a self-defeating task. One of the missions of the ethos of human dignity is to help us overcome this problem. Another task is to bolster our confidence and inspire courage lest excessive preoccupation with vulnerability bring forth despair and cowardice. Human dignity teaches us that moral vision can prevail and that promotion of human dignity is a goal worth striving for, even at the price of considerable risk. Humans are not so vulnerable as to rule out bold action in the promotion of basic human goods.

At the operational level, although it is more respectful of human dignity to educate and wait for people until they interiorize moral values, the vulnerability of potential victims (as well as of the agent who is likely to harm his or her own personality while acting recklessly) prevails upon us to resort to coercion when this is the only reasonable way to protect the basic interests of human individuals. Albeit, precisely because all humans are vulnerable, every act of coercion and every policy of enforcement are always on the edge of pushing people from a position of being potent offenders to that of helpless and hopeless victims. The value of freely made moral choices (dignity as a moral expectation of people) combined with appreciation of human vulnerability bring forth two conclusions: (1) enforcement is necessary for the protection of basic human needs; and (2) mere enforcement is insufficient and sometimes can be quite dangerous (Barilan 2012a).

10.3 Vulnerability as Reflected in Some Rabbinic Writings Concerning Clinical Ethics

In this section I wish to explore a few rabbinic rulings, from which some insights about the rabbinic construction of vulnerability, or at least vulnerability-related questions, can be imported.

The opinions I present here do not necessarily reflect a consensus in the rabbinic world. However, they have all been published by the most authoritative contemporary Orthodox rabbis. Moreover, since there is no central authority in Jewish law (no "official teaching") and since it is customary in Jewish law to follow lenient and permissive opinions in situations of "human distress," many rabbis would endorse such practices *de facto*, even if their personal interpretation of the law is more demanding. Moreover, I will show how these opinions ramify deeply into Rabbinic law and ethics, thus exposing fundamental principles that are at the heart of the consensus in Jewish ethics.

10.4 The Patient who Refuses Necessary Medical Care

In the 1970s the leader of Orthodox Jewry in America was asked what to do in the case of a patient who refuses necessary medical care. In the responsum given, Rabbi Moshe Feinstein considers two possibilities.

If the patient trusts medical judgment but, nevertheless, declines care because of the suffering involved (e.g. side effect of chemotherapy) or due to despair, then care should be imposed as much as possible, as is done with children. The second possibility is that refusal is motivated by mistrust of sound medical opinion, and it is impossible to find a physician who the patient does trust. Then, enforcement of life-saving care should be considered, but only when all of the conditions follow (Feinstein 1985).

The first condition is that "all doctors in the hospital agree [that the treatment is necessary]." The second condition is that the enforcement will not traumatize the patient psychologically, because, as we have seen, in rabbinic psychology, trauma might derange patients in acute conditions. Rabbi Feinstein concludes with a warning against coercive care, even upon the relatives' consent, because imposed care "is not likely to benefit much." The third condition is that the treatment in question is risk-free.

Even if the risk-benefit ratio is clearly on the side of treatment, it is still prohibited to impose risky care on non-consenting patients. The rabbi digresses to discuss the weakness of evidence based medical practice, which in his judgment tends to extrapolate the findings in clinical studies to groups of patients who do not fully match the study groups. He adds that even upon informed consent a patient should not be given a drug whose success rate is less than 50 %.

This responsum was written 20 years before the concept of "evidence based medicine" was borne. From the responsum it is not clear what level of risk to life must be offset by a threshold of 50 % success rate. However, it is worth noting the Rabbi's sophisticated attention to human vulnerability as a holistic, bio-psycho-social concept.

Feinstein acknowledges the vulnerability of both life and self-determination. He notes that people tend to mind present discomfort more than long term benefit, a problem known as "future discounting" (Rachlin 2000). In such circumstances, people tend to bypass their weakness of will by instituting "commitment devices" that preempt non-compliance (Offer 2006). A paradigmatic example is Odysseus' request for his sailors to tie him to the mast of his ship so as to allow him to enjoy the song of the sirens without being able to rush in their direction, as all other seafarers have done and consequently lost their lives (*Odyssey* 12). This psychological mechanism was taken for granted by the Talmudic sages who infer the right of the government to levy taxes (Exodus, 25) for the administration of necessary but very unpleasant medical care. It is like a physician who ties a patient to a tree in order to cauterize a foot [ulcer]. (TOSEPHTA *Sheqalim* 1, 6).

Feinstein mentions people's tendency to over-value the moment, but since future discounting is the chief cause for weakness of will (Ainslie 2001), my reading into this comment may suggest that informed consent to care, as expressed in acceptance of therapeutic relationships, entails implicit permission to impose care in moments that are particularly vulnerable to weakness of will. However, in line with contemporary sensibilities on vulnerability, Feinstein acknowledges autonomy as a specially valued and vulnerable locus of human dignity. Odysseus' self-imposed inability to undo his shackles and the patient's inability to untie himself from the tree are restrictions that act out their autonomy, and do not diminish it. This is a unique feature of the vulnerability of human dignity, whose prosperity depends on measures of anticipatory self-restrictions of free choice. As we can see, the rabbis took the practice of patients' self-binding for granted, and inferred from this context the possibility of self-binding in non-medical contexts.

Suppose we are in doubt which moment better reflects one's authentic autonomy—the desire to keep the ship safe or, perhaps a few moments of proximity to the music of the sirens. Perhaps, as some romantics might suggest, instead of smothering the desire to rush after the sirens, we had better overcome fear of death and attachment to life for the sake of celestial beauty and supreme pleasure. Indeed, as Donald Davidson explains, humans need a system to prioritize values, methods for valuing values in order to explain why life is more important than a few moments of celestial music or a few moments of suffering (Davidson 1970). We find out that our vulnerabilities, and the vulnerability of one of autonomy's and identity's most central feature—free will—prevail upon us to engage in second order moral valuation (i.e. the valuation of values relative to each other) and the projection of values over long segments of time, all the way to spiritual pondering of the most basic and ultimate values, and the capacity to envelope whole lives with meaning. Arguably, this is the eventual fulfillment of autonomy. In the absence of vulnerability, we would not be motivated to develop our autonomy and identity so much.

While despair and fear of suffering do not embody autonomy, sincere personal discretion in the service of a moral goal is the hallmark of autonomy. Autonomous judgment is expected to stand firmly against social pressures to conform. But autonomous judgment might be mistaken as well, as the doctors and the family firmly believe. Feinstein constructs his responsum so as to meet this very dilemma of apparently clashing judgments.

On the one hand he acknowledges the rational decision to save life by means of proper medical care, and not to lose life due to misjudgment and problems of trust. On the other hand, Feinstein also reminds the caring person who presented the question that professional judgment is vulnerable as well. With tongue in cheek he points out that consensus among experts is a rarity, and that even cutting edge and scientifically based medical knowledge cannot capture fully and confidently the situations of particular patients. The thrust of Feinstein's opinion is borne out by his presumption that coercion, even by the most competent and benevolent persons, is likely to be self-defeating.

From a scientific perspective, it is not clear at all that imposition of care must be fatal, even indirectly; but bioethical sensibilities hold respect for autonomy as no less valuable than life itself. I am not sure what exactly Feinstein had in mind, but we can see his practical wisdom in this responsum. He manages to stay loyal to traditional moral values, reiterating pre-modern rabbinic opinions that life is more valuable than personal choices to die or to avoid suffering. But he gently neutralizes this maxim by stepwise invocation of human vulnerabilities and the vulnerability that results from their interconnectedness. Lastly, as an Orthodox rabbi holding fast to traditional ways of life, one would expect Feinstein to promote "community values" and the role of the family and professional authority in care and coping. But Feinstein is attuned to the destructive potential inhering in good, well-intentioned and devoted intimate care and "family values."

Rather than bolstering the determination of family and doctors who know the patient and are sincerely committed to his wellbeing, Feinstein asks them to reflect on the vulnerability of autonomy, particularly when life is at risk and the patient fails to find sympathy and understanding for his own judgment. Feinstein refrains from reliance on his authority in order to pronounce an unambiguous "pro-life" statement. Valuation of life and responsiveness to vulnerability do not permit the simplicity of straightforward judgments.

The 1996 Israeli Patients' Rights Act empowers institutional ethical committees to enforce life-saving emergency treatments whenever a committee believes the patient is likely to be grateful upon recovery (Siegal and Gaitini 2002). This unique clause in the law was originally written by Rabbi Mordechai Halperin, expert on Jewish law and medical ethics. Many ethicists in Israel and abroad criticize it as paternalistic. In my view it is a humane and empowering clause, a kind of commitment device helping us to cope with our own vulnerabilities.

10.4.1 The Mildly Retarded Woman who Refuses Cesarean Surgery

In another case an Israeli woman was expecting a child whose medical condition was an indication to cesarean surgery. The doctors opined that the baby was not likely to survive the physiological stress of natural birth. Alas, the woman refused to grant consent to surgery, notwithstanding the lethal risk to the child. She simply stated again and again that she found such surgery too frightening for her. The case was brought post-factum to the attention of Rabbi Auerbach, whose authority in Israel paralleled that of Feinstein in America. Auerbach's opinion was against coercive surgery, arguing that a person has no duty to subject herself to risk and fright for the sake of saving another (Avraham 1997, *Orah Haim*, Mark 330). From the text it seems that the woman had borderline mental retardation or a mental problem, barring her from processing fully the risks and benefits.

It would be erroneous to talk about the vulnerability of her autonomy. However, absence of mature capacity for autonomous judgment is also a locus of vulnerability. While a competent person may resent the imposition of care, he or she is also capable of understanding the situation from the perspective of caregivers. The woman in question is not likely to reach this level of integrative understanding. She might suffer the terror of coerced anesthesia and recovery from surgery as a mysterious and sinister invasion of her body. Her restricted capacity either to understand or cope with the situation renders her even more susceptible to the trauma of forced care, because she will experience terror and pain without understanding the procedure and its justification.

Although the unborn do not have the status of legal persons in Jewish law (Barilan 2009/2010), Auerbach does not resort to this line of reasoning. Although he stresses the duty to exhaust all means possible in order to persuade that woman, he does not find a legal way to force surgery on one person for the sake of saving the life of another. Perhaps he was confident that painstaking efforts at counseling and persuasion would have worked out.

Apparently, the Israeli secular court was less sensitive to the three aspects of vulnerability which Auerbach addressed: (1) the social vulnerability of a woman with cognitive difficulties; (2) the anguish of physical suffering due to forced violation of the integrity of the body; and (3) the situation of a pregnant woman whose psychological and even medical wellbeing are often subordinated to the life of their unborn children. The fetus is vulnerable as well, of course. Yet, in Jewish law there is a clear preference for the vulnerability of the mother even at the expense of the fetus. No rational argument is propounded in support of this preference. One prominent medieval authority, *Rashi* (France, eleventh century) attributes this to the positive Divine law, according to which the unborn is not a legal person (gloss on *Talmud Sanhedrin 72b*); another, no less authoritative Medieval authority, Maimonides (twelfth century, Egypt) explains this by the "naturalness" of self-defense (*Mishneh Torah, Hilchoth Rotzei'ah* 1:9).[1]

[1] Maimonides does not use the concept of "nature" which was not known in Jewish law and thought. He writes that "this is the way of the world (כך דרכו של עולם)." It is not clear that he meant to say that killing aggressors, even when innocent, is "natural" [whenever this is the only reasonable means to divert the lethality of the aggression]. Although Thomas Aquinas does not discuss life-saving abortions, he elaborates on the permissibility of self-defense (Summa Theologica, II-II q.64.a7). Thomas constructs it on three moral arguments – the "principle of double effect" (actually, only later would Thomas' words be recognized as constitutive of the "principle"), the "natural" drive of self-preservation, and Christian "Charity" (in 22 64.6 he explains that charity compels the duty to protect one's own life. See also Renick 1994). It follows that both Maimonides and Thomas Aquinas permit some form of direct killing by the invocation of "the natural" and by a very peculiar, even counter-intuitive conceptualization of "charity" or "mercy." However, whereas Maimonides explicitly permits lethal self-defense even against innocent threats (e.g. the unborn child who poses a risk to the life of the mother), Thomas's position on this very point is unclear. Thomas requires that the death of the aggressor be "outside the intention" of the defender, who should only think about the reasonable force required to repel the lethal attack. Whether awareness of the culpability or innocence of the aggressor should be within the awareness of the defender is an open question. See also Barilan 2012a pp. 212–220.

Moreover, the very perception of the unborn as innocent and helpless adds weight to the pressure on women to self-sacrifice their health interests and integrity of body and person for the sake of their own children. Hence, in a unique ruling, Maimonides writes that Jewish law prohibits having mercy for the unborn when the life of the woman is at risk (*id*).

Although Catholic and Jewish teachings on abortion and the moral status of the unborn are considerably different, I have found a Catholic teaching similar to Auerbach's. A manual for hospital staffs, published with the authorization of the Church (*Imprimatur*), writes,

> Even when the section [i.e. cesarean surgery] is permitted, it is to be questioned whether a dying mother is morally obliged, practically speaking, to suffer the operation. In this instance surgery may with caution and gentleness be suggested to her, but any mention of a moral obligation to provide surgery for the Baptism of the child [in those days it was not possible to tell whether the child was viable or not, but it was assumed that is would live at least a few seconds so as to allow for baptism—*author*.] is to be prudently avoided. A dying mother is in no condition to ponder such an obligation (Bowen 1932, p. 8).

When these words were printed, mortality from cesarean surgery in the USA was in the range of 4–16% (Plass 1932). One can hardly imagine more vulnerable humans than a child and mother who are entrapped in mortal danger during childbirth. The child is not capable of choice; confronting the mother with choice might exacerbate her anguish and misery. Autonomy is at the heart of vulnerability; but certain vulnerabilities render it more respectful of the person to avoid autonomous choice, than to seek it at the moment of crisis.

10.5 Trust and Vulnerability

The sixteenth century codex of Jewish religious law, *Shulkhan Aruch* (YORE DE'A 336:1) does not write that the sick have a duty to seek and receive medical care. This is not because Rabbi Karo, the author of the codex, does not think that people have a religious and moral duty of self-care. Here are his words about a person who refuses charity money that is needed to save his life:

> Whoever has to take charity [money] in order to survive, such as the old, the sick and the sufferers, and refuses it, is like a spiller of blood, and his sorrows [are not heroic or sacrificial] but only sins. He deserves death [in the hands of God]. (Yore De'ah 255:2).

The reader might wonder why such language is not employed with regard to a patient who refuses life-saving care for other, non-pecuniary, reasons. An answer may be found, in my opinion, in Karo's choice to explicate the deontology of care from the side of the care-giver,

> …A [physician] who abstains from caring for a patient is like a spiller of blood. Even if other physicians are available for that patient [the physician must not decline a request to care], because a patient may not benefit from every physician (*Yore De'a*.).

Money distributed by the welfare authorities is never tainted. It raises no questions of trust either. Everybody knows what the value of money is. So if a patient seeks food or a certain therapy, there is no excuse for not using money in order to procure them. On the other hand, one may have all the money in the world at one's disposal, and still feel distrustful of the therapy offered to him, or be reluctant to endure its pain. Nobody has a religious duty to become a patient. Only those who choose to be patients have an unequivocal duty to exert every effort, and even spend all their property on care.

The vulnerability of trust does not allow its invocation in the language of duty. Neither does it allow coercion. Like love, hope and faith, trust is a relational attitude that cannot be produced mechanically. Ultimately, our most basic values such as survival, care and continuity (procreative sex) depend on the good will, free choice and mutual trust of vulnerable people, whose vulnerability cannot withstand the burden of dogmatic laws and coercive measures. Paradoxically, this very vulnerability creates even stricter duties for the less vulnerable. While in Western medical ethics a physician has no duty to accept patients who have other doctors to mind them, Jewish law does not permit a physician to reject a patient who sincerely believes that it is better for him or her to be cared for by a particular doctor. Evidently, Jewish law does not expect doctors to undertake unreasonable load of responsibilities, but it expresses faith in the relative resilience of the care-giver to be able to try a little harder and see one more patient.

The extreme harshness of some life and death dilemmas and the ambiguity of more ordinary problems of vulnerability have brought into light the limits of explicit bioethics, of formal, legal and deontological modes of normativity. Our first response to vulnerability is to defend humans, especially the vulnerable, by means of law and law enforcement. But Jewish law hints that this might be a self-defeating conceit. So is the conceit of realizing our autonomy without acknowledgment of our vulnerabilities, especially weakness of will and distorted judgments in acute moments of crises. Either way we decide to take vulnerability into consideration, we depend on trust, care, hope and similar states of mind that are beyond coercion and manipulation (Barilan 2012b). They are also somehow mysterious, as we have only partial insight into, understanding of, and power over, these cherished and vulnerable attitudes and virtues.

Acknowledgment This research was supported in part by a grant from The Israeli Scientific Foundation (ISF 197/10). This work is dedicated to my youngest child, Mario Benjamin, born only a few months ago, my vulnerable strength—בנימין

References

Ainslie, G. 2001. *Breakdown of will*. Cambridge: Cambridge University Press.

Albrecht, G. L., and P. J. Devlieger. 1999. The disability paradox: High quality of life against all odds. *Social Sciences and Medicine* 48:977–988.

Avraham, A. S. 1997. *Nishmat Aavrham*, 5 vols. Jerusalem: Falk-Schlesinger Institute. (Hebrew).

Bach, J. R., and M. C. Tilton. 1994. Life satisfaction and well-being measures in ventilator assisted individuals with traumatic tetraplegia. *Archives of Physical Medicine and Rehabilitation* 75:626–632.
Baker, K. K. 1997. Once a rapist? Motivational evidence and relevancy in rape law. *Harvard Law Review* 110:563–624.
Barialn, Y. M. 2009/2010. Her pain prevails and her judgment respect: Abortion in Judaism. *Journal of Law and Religion* 25:97–186.
Barilan, Y. M. 2012a. *Human dignity, human rights and responsibility: The new language of global bioethics and bio-law*. Cambridge: MIT Press.
Barilan, Y. M. 2012b. From hope in palliative care to hope as a virtue and a life skill. *Philosophy, Psychiatry and Psychology* 19:165–181.
Barilan, Y. M. 2014. *Jewish bioethics: Rabbinic law and theology in their social and historical context*. Cambridge: Cambridge University Press.
Bok, D. 2010. *The politics of happiness: What governments should learn from the new research on well-being*. Princeton: Princeton University Press.
Bowen, J. R. 1932. *Baptism of the infant and fetus: An outline for the use of doctors and nurses*. Dubuqe: M.J. Knippel.
Darley, J. M., and C. D. Batson. 1973. From Jerusalem to Jericho: A study of situational and dispositional variables in helping behavior. *Journal of Personality and Social Psychology* 27:100–108.
Davidson, D. 1970. How is weakness of the will possible? In L. Foster and J. Swanson (Eds.), Experience and Theory. Amherst: University of Massachusetts Press.
Feinstein M. 1985. Responsa "Igrot Moshe," vol. 7. New York.
Haidt, J. 2001. The emotional dog and its rational tail: A social intuitionist approach to moral judgment. *Psychological Review* 108:814–834.
Isen, A. M., and P. F. Levin. 1972. Effect of feeling good on helping. *Journal of Personality and Social Psychology* 21:384–388.
Kavka, G. 1983. Hobbes's war of all against all. *Ethics* 93:291–310.
Lakoff, G., and M. Johnson. 1999. *Philosophy in the flesh: The embodied mind and its challenge to western thought*. New York: Basic Books.
Lamont, E. B., and N. A. Christakis. 2001. Prognostic disclosure to patients with cancer at the end of life. *Annals of Internal Medicine* 134:1096–1105.
Offer, A. 2006. *The challenge of affluence: Self-control and well-being in the United States and Britain since 1950*. Oxford: Oxford University Press.
Plass, E. D. 1932. *Forceps and cesarean section. White house conference on child health and protection*, 215–247, New York: The Century.
Rachlin, H. 2000. *The science of self-control*. Cambridge: Harvard University Press.
Renick, T. M. 1994. Charity lost: Secularization and the principle of double effect in the just war tradition. *The Thomist* 58:426–441.
Siegal, G., and L. A. Gaitini. 2002. Treating acute anemia in a Jehovah's witness in Israel: An innovative approach to a medical and legal challenge. *Medicine and Law* 21:485–492.

Part III
Responses to Vulnerable Groups from Six Religions

Chapter 11
Buddhist Perspective on Four Vulnerable Groups: Children, Women, the Elderly and the Disabled

Soraj Hongladarom

11.1 Introduction

In bioethics, the concept of human vulnerability is commonly used in the context of research in biomedical sciences where research has the potential, not only of producing real and lasting benefits to human society as a whole, but also the potential to harm those participating in the research in a variety of ways. Furthermore, the concept also highlights the frailty of human beings, their susceptibility to a variety of harms and injuries simply by virtue of their being human beings. In this sense the concept acts as a counterpoise to that of autonomy, which emphasizes the individuality of human persons and their capabilities in making their own decisions.

While the concept of autonomy is often based on the idea that humans are individualistic, each of whom are "ends in themselves" reigning in their own separate, private domain, the concept of vulnerability emphasizes interlinking among human beings and their solidarity. Since all of us are vulnerable, it is necessary for us to help and take care of one another. This dimension of caring is obviously important in bioethics, but seems to be rather neglected when the dominant discourse is centered on the concept of autonomy.

In talking about vulnerability, one can separate talks about human vulnerability as a whole, or about the vulnerability of human beings as a species, from those about certain groups of humans who are more vulnerable than others. This paper will focus on the latter kind of talking, focusing on how Buddhism looks at these vulnerable groups, namely children, women, the elderly and the disabled. These groups are clearly vulnerable because they are weaker, both in the physical sense, and also in the social sense of being less advantaged, than the dominant group. But the problem does not limit itself here. Among these groups themselves there are issues that cut across them such that there are degrees of vulnerability within each group too.

S. Hongladarom (✉)
Department of Philosophy and Center for Ethics of Science and Technology, Chulalongkorn University, Bangkok, Thailand
e-mail: Soraj.H@chula.ac.th

J. Tham et al. (eds.), *Religious Perspectives on Human Vulnerability in Bioethics*, Advancing Global Bioethics 2, DOI 10.1007/978-94-017-8736-9_11,

For example, there are more socially advantaged women, such as those who enjoy more education and more income than their counterparts in the same society. So it is clear that the women who have these privileges are less vulnerable. This shows that vulnerability is not an issue of physical properties (whether one is a woman, a child, or is disabled), but has an important social dimension. I would like to analyze the concept of vulnerability further and I propose to use insights obtained from the Buddhist teachings to do so. The main questions of the paper, then, are: What is the perspective of Buddhism on human vulnerability? What kind of justification or reasons does Buddhism have in support of the idea that the vulnerable should be protected?

The latter question is important because the mainstream position in bioethics in the West today appears to be one that relies on the concept of individual rights based on the language of autonomy. Since the vulnerable also have their own rights by virtue of their being humans, their needs should be protected. This line of reasoning, however, does not differentiate between the vulnerable and those who are not. Perhaps the argument is such that the vulnerable have a right to be treated differently than the non-vulnerable because doing so would restore normal functions to the vulnerable so that they could become fully participating citizens in the "kingdom of ends," to use Kant's term. Based on the Buddhist teachings, I would like instead to argue that the language of rights and autonomy is too demanding and legalistic, and in many cases would not be as effective in promoting the welfare of the vulnerable as the language of interdependence and compassion.

One talks of interdependence and compassion when one finds that there is an essential link that binds up all human beings with one another, so that each one of us is in some way one and the same. Hence promoting the welfare of the vulnerable, recognizing them for what they are and reducing the conditions that have led them to become disadvantaged are what we should be doing as beings who are compassionate toward one another. This view of beings who are compassionate to one another is based on the Buddhist metaphysical view of the interdependence of all human beings, indeed all beings, on one another. I will show how looking at our fellow human beings, vulnerable or not, in this way is more effective in realizing the goals of having the concept of vulnerability in the first place than the right-based way.

11.2 Concept of Human Vulnerability

The whole concept of vulnerability itself is easy enough to understand. The word 'vulnerable' originally means 'easily wounded.' It comes from Latin '*vulnerabilis*' meaning 'wounding' or 'injurious.' '*Vulnerare*' is a verb meaning 'to wound' and '*vulnus*' means 'wound.' Thus one who is vulnerable is easily wounded; one has a picture of a soldier who is injured in battle but continues fighting. He is vulnerable, especially at the location of his wound.

His vulnerability exposes him to the enemies who certainly would target his wound as a way of easily defeating him. A classic example, of course, is Achilles, who is vulnerable at his heel. The vulnerability distinguishes him from another soldier who is not wounded because his wounds make him less able to fight alongside those who are fully healthy. In the same vein, we also say of a city that it is vulnerable when there are some weak spots in its defense that makes it easy for the enemy to attack. Perhaps a wall at one side of the city is weaker than those at other sides, so if the enemy knows this they can target this wall which could result in the whole city falling down. Here one does not mean that the city has a wound, but is making a metaphorical comparison between the city and the body of a soldier. When the soldier is vulnerable, he can become an easy target for attack. So is the city when it is vulnerable.

The term 'vulnerability,' then, when used in the military context refers to a situation where the defense is weaker at one spot or another, making it exposed to the enemy attack. Any kind of defense, then, is liable to being vulnerable. We would say that a tank is vulnerable when its armor is too thin to withstand enemy fire, for example. Furthermore, one can also use the term in the context of attack, such as when one says that an attack is vulnerable because it contains some weakness that would result in its failure. Thus, one sees that the term is used in a situation when either an attack or a defense contains some weak spots that would result in failure in achieving the military objective.

The term is also being used outside the military context. One says of someone's "vulnerable reputation," meaning that he has some weak spot in his reputation that could result in his reputation being damaged. Hence, generally speaking, one uses 'vulnerable' when one wants to refer to any type of situation where there are some weak points that could result in being invaded or attacked resulting in damage, loss or failure of the whole thing. It is interesting to see how much this original sense of 'vulnerability' is retained when the context is as complicated as that of biomedical science and research. In that context, a person is vulnerable when he or she contains some weak spots, so to speak, that would result in he or she being injured or suffering from loss of dignity or worse. Usually children are considered especially vulnerable, because they are small and are incapable of defending themselves and retaining their autonomy vis-à-vis the adults.

Thus it is universally acknowledged that children deserve special care and protection when it comes to biomedical research. Other groups, such as women and the elderly, are in the same general situation.

It seems, then, that the concept of vulnerability should be clear enough when it is transported to the context of biomedical science and research. However, there are many debates and controversies in the literature on bioethics about precisely how the concept is to be understood. These controversies are focused on how the general meaning could be translated into finer shades of meanings which reflect a wide variety of viewpoints. For example, Ruth Macklin offers a definition of a vulnerable person as one who can be exploited easily, and adds that neglecting the vulnerable persons is morally wrong because exploitation is morally wrong. The problem for Macklin then is how to play up the concept of exploitation with that of vulnerability.

She sees that one can be exploited without being oneself vulnerable to harm, and one can also be harmed without being exploited. Thus Macklin tries to show that defining vulnerability through the concept of exploitation does not work because there are cases of there being the former but not the latter (Macklin 2003). However, Macklin's focus on harm does not seem to do away with the difficulty here, for one can also imagine another situation where one is not harmed but still vulnerable, such as when someone is in a precarious situation even though no actual harm to her has occurred. Furthermore, the harm can happen at many levels, such as the physical, mental, or psychological level, and not only individuals can be harmed, but entire groups as well. When the structure of a society is such that an entire group is disadvantaged and underprivileged, then it can be said that the group is vulnerable.

Perhaps the original document in research ethics on vulnerability is the Helsinki Declaration, which states as follows:

> Some research populations are vulnerable and need special protection. The particular needs of the economically and medically disadvantaged must be recognized. Special attention is also required for those who cannot give or refuse consent for themselves, for those who may be subject to giving consent under duress, for those who will not benefit personally from the research and for those for whom the research is combined with care.[1]

The Helsinki Declaration makes clear it that those who are vulnerable need special protection. What is interesting here is that the passage gives us some idea of how the vulnerable is to be recognized. First of all they are economically and medically disadvantaged. This of course leaves many details unmentioned, for example how much disadvantaged a population should be in order to be qualified as vulnerable, how 'medical disadvantage' should be defined. If someone has a tendency to get diabetes because of her genetic makeup, does that qualify as a medical disadvantage? Who should decide whether a population or an individual has a medical disadvantage or not? These are very thorny issues, and a lot of sustained effort and discussions are needed to sort them out.

The passage from the Helsinki Declaration also tells us some of the conditions for someone or some population group to become vulnerable. Apart from being disadvantaged economically and medically, they also include those who "cannot give or refuse consent for themselves." This condition also has raised a number of concerns. For example, there might be those who are able to give or refuse consent for themselves, but their conditions are such that they are vulnerable nonetheless; hence they should also be protected. The elderly or the poor who are still able to think and decide for themselves may, on the surface at least, be able to give or refuse consent, but their situation is such that, by the very nature of their being elderly or economically poor, they might need special protection.

For the elderly, their physical constitution appears to be in need of special care; hence since much of the research in advanced biomedical sciences today involve intervention and risks to the bodies of the participants, then the elderly need more protection simply because of their being elderly. This should be the case even if the elderly in question are fully capable of making autonomous decisions. As for the poor, there are many discussions about them being induced to participate in the

[1] World Medical Association (1964). Declaration of Helsinki, as amended by the WMA 52nd General Assembly, Edinburgh, Scotland, October 2000. Paragraph 8. Quoted in Macklin (2003).

research because of their economic needs. We will talk more about this later, but here the issue is that biomedical research should at least contribute to a fairer world where there is less inequity and injustice. Thus, simply on this principle alone the economically poor need some special protection even if they are fully able to make decisions for themselves.

At the very least, the research should not exacerbate existing patterns of economic exploitation and injustice that abound in the world. Ethics for researchers should not be limited only to how the researchers interact with their participants alone, but should take need of the overall social and economic context in which the research is done too.

Another of the conditions mentioned in the Helsinki Declaration is that those are vulnerable who "may be subject to giving consent under duress." Again questions concern how to tell when the participant is actually subject to giving consent under duress. Of course we can think of extreme cases, such as the researcher pointing a gun to a potential participant's head demanding the latter to consent to the research. But in almost all cases that does not happen; on the contrary, the "duress" in question here can be very subtle. Many discussions and debates are centered on whether giving inducements qualify as putting the potential participants under duress or not.

In normal, non-research situations, putting someone under duress is not the same as providing him or her with inducements. If I demand something from you and threaten to harm you physically, that is duress, but if I ask you to do something for me and promise to give something in return, that would not count as duress under normal circumstances. However, in the context of biomedical research providing inducements is generally prohibited because that would jeopardize the principle of autonomy. But in normal circumstances, that is in other contexts outside of biomedical research, someone has full ability of making autonomous decisions even when there are "inducements"—I am the one who decide whether I should take the so-called inducements or not. At any rate we are seeing here that the words of the Helsinki Declaration need to be spelled out clearly in order for them to work fully. That would require much more thinking and deliberation.

The next to last condition, namely "those who will not benefit personally from the research" could somehow still be qualified as "vulnerable" if not benefiting from the research counts as an instance of inequity or injustice. Hence those who do not get the benefits are vulnerable because they are shortchanged and taken advantage of while other groups obviously do benefit from the successful research. Here the concept of vulnerability seems to be extended to include those who do not have any real wounds at all, either physical, psychological or social, but they are included in the scope of the concept because their metaphorical "wound" is their very position which excluded them from benefiting from the research.

The last condition mentioned in the Helsinki Declaration concerns those "for whom the research is combined with care." This is very important and it points to the unequal relation between the doctor and her patient. Here the vulnerability lies in the patient's lack of power or in her position as subordinate to the doctor, which makes the patient exposed to being wounded by the doctor who in this case is doubling as a researcher. Many ethical guidelines pay special attention to therapeutic misconception as a serious threat. Here also there are deeper shades of meaning that need to be spelled out later on.

For example, how can the patient make a fully informed and autonomous decision when she is lying in the hands of the doctor who is also conducting research in addition to providing the medical care? If the topic of research happens to be the same as the illness that the patient is suffering, then the issues become harder to distinguish. One might also question the presupposition behind this guideline, which seems to be that research and care should be clearly separated. But perhaps one could imagine a situation where the caregiver and the researcher, who is embodied by one person, is treating a patient and conducting research and, out of compassion, steers the research solely for the purpose of finding the best care for the patient. Here the research and the care become one and the same.

This scenario does not happen much anymore when biomedical research has become very capital-intensive and multi-national, but the possibility of combining research and care shows that the vulnerability does not come automatically from combining research and care together.

The concept of vulnerability has become prominent in recent years as a counterweight to the concept of autonomy. A standard view in mainstream bioethics has been that a person is autonomous, and is only vulnerable when certain conditions occur that prevent the person from functioning normally. However, viewing a person only as an autonomous agent seems to miss an important aspect of a human being, which is that vulnerability does pervade the very condition of being a human. A human is susceptible to disease, is in a frail condition relative to the environment, and so on. As a vulnerable person deserves and needs active protection, categorizing every human being as vulnerable then can function as a guiding principle in bioethics where each human being is accorded a special status that emphasizes the need for protection and special care.

This dimension, however, appears to be lacking in the normal characterization of humans only as autonomous agents.

In short, there are two aspects of human vulnerability.[2] One is that every human being is vulnerable simply by virtue of being a human. The other is that one is vulnerable in this sense only when one happens to be in a certain condition that prevents one from being able to function or perform as well as a 'normal' person. These two aspects can together provide a foundation for a number of bioethical guidelines that emphasize, not each individual taken in isolation, but human solidarity and capacities that each human has in relating to another, helping and caring for one another.

[2] According to Solbakk, the two aspects of human vulnerability, one that emphasizes the vulnerability of humans a whole and one that focus on particular human groups relative to the mainstream one, both appear in the text of the UNESCO Declaration of Bioethics and Human Rights as a result of a compromise. However, both aspects are equally important in highlighting the importance of the concept of human vulnerability as a whole as a counterpoise to the dominant paradigm of autonomy (Solbakk 2011).

11.3 The Buddhist Position

So what is the Buddhist position in all of this? Two key ideas in Buddhist philosophy are those of interdependence and compassion. Interdependence means that everything in nature does depend on another for its very being. A thing is what it is because it is connected and related to all other things in nature. The reason why a table is what it is is due to many factors, such as its shape and function. The shape of a thing is defined through the relation of the thing with others surrounding it; if there were no such relation then that thing would not be a thing at all because one cannot find the edge beyond which that thing ceases to be that thing.

Furthermore, even if a thing may have a shape of a table, but if it is not used *as* a table then one could say that it is not a table at all. If a big, wooden object which looks like a table is used as, say, fuel, then it becomes fuel, at least in the conception of those who would like to burn it, rather than remaining a table. So whether the thing is a table or fuel depends ultimately on how it is being used or going to be used. Either way its being a table (or fuel) depends on its function, its role in relation to those who are using it. Hence, things are what they are because of their relations to other things, and according to Buddhists this is the case at all levels, from the huge macro-object to the tiniest micro-object.

Another idea is that of compassion. In Buddhism this is arguably the most important element. It is the essential quality of the Bodhisattvas, or those who dedicate themselves to ultimately becoming a Buddha, an Enlightened One, so that they are able to help ferry sentient beings across the sea of suffering called *samsara* to the shore of Liberation. Compassion is the feeling one has when one feels that the sufferings of others are one's own. The term 'compassion' in English is the usual translation of the Sanskrit *karuna*, meaning the wish that all beings are relieved of suffering.

When we perceive somebody to be suffering, for example when we enter a hospital and see a number of patients are who suffering from a large variety of illnesses, we feel compassion when we sincerely and genuinely feel that their sufferings should go away and wish them to get rid of all sufferings of any kind; and we are willing to do anything in order to relieve them of suffering. Interdependence is the basis for compassion because when one sees that things are all interdependent, one then realizes that one's own being is connected with those of others. The connection can be so deep as to see that the very being of oneself would not be there if not for the being of others, in effect seeing that oneself is in others and that others are in oneself. Seeing things that way the bodhisattva then realizes that the sufferings of others are in fact his own suffering. Since everybody is essentially connected, the pains and sufferings felt by one cannot be entirely alienated to that person alone, but must be shared by others.

This sharing is not the mere feeling of sympathy, such as when one feels sorry for others who are less fortunate and who are suffering. On the contrary it is an objective understanding that the sufferings felt by others are in fact one's own, since the bounds that separate oneself from others, are ultimately speaking, an illusion from the beginning.

We don't have to be bodhisattvas or liberated beings in order to become deeply compassionate beings who sees the objective interconnection of all beings with one another. In any case, the Buddhist viewpoint on human vulnerability stems directly from the understanding of interdependence and compassion here. Since everyone is dependent on everyone else, everyone deserves compassion from all others. And if this is the case, then the vulnerable actually need more compassion than others because of their precarious situation. As mentioned before, the compassion here is not limited only to the feeling one may have toward others who are less fortunate, but essentially includes *active* aspect where doing things for others so as to relieve their sufferings count very much as compassion too.

Here one finds the main difference between the Western, right-based approach and the Buddhist one presented here. In the former the reason why we should actively care for the vulnerable is because they have rights, which demand that others have duties to care for them. Their rights are based on the idea that they are fully capable human beings, who are made vulnerable because of external circumstances. Thus their status as being vulnerable does not detract from their status as fully functioning members of the kingdom of ends. Their individual status as subjects who are capable of understanding and making autonomous decisions entitles them to be subjects of rights, which means that others must act in one way or another in order to meet their needs.

This view has incurred some philosophical difficulties since it is based on the idea of autonomous decision making. But many are vulnerable precisely because they are incapable of making such decisions. The ideal is that one who is capable of making autonomous decisions is accorded the status of the subject of rights, but there are, for example, those suffering from dementia or other forms of illness which render them incapable of making autonomous decisions, or any decisions at all for that matter. In this case, the right-based approach would say that these people are still subjects of rights because of their default status as capable human beings. That is, they used to be fully functioning human beings before and their suffering from dementia is a condition that happens to them, something added on to them which does not alter their original status.

This point would be clearer if the comparison were instead made with another kind of being who is not capable of making any autonomous decisions from the beginning, or as part of their biological nature. This is of course why animal rights are not recognized as enjoying the same legal protection as human rights. This way of looking at rights is based on the metaphysical idea of individuals who are separate one from another. The ultimate reference point is that of the autonomous subject. This presupposes that the subjects are metaphysically constituted in such a way that any interconnection among them is not emphasized. This is in contrast to the Buddhist position discussed above.

In Buddhism, it is the interconnection or interdependence, not the status as autonomous subject, that is the ultimate reference point of ethical consideration. And the consideration is based in Buddhism on compassion rather than on performing duties as in the Kantian position.

In this sense, then, thc vulnerable are, according to the Buddhist perspective, those who need special protection and care due to their conditions which make them more susceptible to harms. Giving them special attention and more care is justified because they are weaker and are hurt more easily. The harm here is not limited to physical, but includes social harms such as social injustice and inequity.

Thus the vulnerable also include those who are on the "wrong side," so to speak, of the justice scale, so that something needs to be done about them for justice to be restored. According to the Buddhist perspective, the vulnerable need more compassion because of their special position.

It is true that in fact everybody needs compassion because everyone is our fellow being, all swimming in the sea of suffering or *samsara*, but still the vulnerable need more compassion, or more action directed at helping them, because they are more likely to be harmed. For example, one who is physically injured needs some special attention to dress and treat his wound and so on. He is vulnerable because of the injury and, while it is true that one should be compassionate to everybody, it is also true that one should be doing something extra in order to help this particular person who is injured so that his wound is healed. I believe that this point underlies the Buddhist viewpoint toward the vulnerable.

In what follows I will discuss the four main vulnerable groups, namely children, women, the elderly and the disabled, then discuss how these groups are viewed in Buddhist thought and how their special needs should be addressed.

11.3.1 Children

Children are especially vulnerable precisely because they are incapable of defending themselves and are in need of protection by adults. Moreover, they are not yet capable of making autonomous decisions so as to offer informed consent on their own, nor enter into any contractual obligation for that matter.

Thus children are usually protected very strongly against being subject to research. This, however, has led to a reverse situation where research which would lead to material and health benefits to children themselves are not conducted adequately due to the stringent conditions required for doing research on children.[3] Nonetheless, it is clear that children as a group demand special care and protection from exploitation because, if left to themselves, they are powerless against those adults who might want to make use of them in one way or another. As adults are obliged to protect children even if the children are not their own offspring, so too must the research community protect children as a group because of their vulnerable status. It is only in certain, specifically defined situations where the research has clearly demonstrated potential to benefit children that research is allowed to be

[3] Doriane Lambelet Coleman has outlined what she calls a "prevailing view" which holds that healthy children should be included in research in order to share the burden with non-healthy ones, a position with which she strongly disagrees. See Coleman (2007). The so-called prevailing view, according to Coleman, can be found in Kahn et al. (1998).

conducted on them. In this situation special provisions and conditions have been designed to make sure that their vulnerable status is not being taken advantage of.

Some international documents on research ethics do mention children specifically as an example of vulnerable groups needing special care and protection. For example, the UNESCO Universal Declaration of Bioethics and Human Rights says the following as part of Article 14:

> 2. Taking into account that the enjoyment of the highest attainable standard of health is one of the fundamental rights of every human being without distinction of race, religion, political belief, economic or social condition, progress in science and technology should advance:
> (a) access to quality health care and essential medicines, especially for the health of women and children, because health is essential to life itself and must be considered to be a social and human good; (UNESCO 2005)

The purpose of this Article in the Declaration is to integrate scientific and technological developments into justice and equity in general, and health care in particular. Progress in science and technology should promote access to health care and essential medicines, especially for the health of women and children. Hence UNESCO declares that access to health care and essential medicines are human rights and states should make sure that these basic health needs are met, especially for women and children. The reason why these two groups are singled out is because they are vulnerable and are weaker relative to the dominant group, which is that of adult males. So states need to ensure justice and equity for all the groups in their domains. In order to do that sometimes it might become necessary to allocate some resources to these vulnerable groups first.

Furthermore, the CIOMS guidelines for researching on children are more specific:

Before undertaking research involving children, the investigator must ensure that:

- the research might not equally well be carried out with adults
- the purpose of the research is to obtain knowledge relevant to the health needs of children
- a parent or legal representative of each child has given permission
- the agreement (assent) of each child has been obtained to the extent of the child's capabilities
- a child's refusal to participate or continue in the research will be respected (CIOMS 2002)

The purpose of these guidelines is to ensure that children as a vulnerable group are not being taken advantage of during their participation in biomedical research. Note that instead of informed consent what the guidelines are calling for is "agreement" or "assent" from the child, presumably in verbal form.

The research, however, needs to make sure that the child understands what is going on, to the extent he or she is capable of doing so. All this needs to proceed out of genuine care and compassion for the child. Researchers should not merely follow the letter of the guidelines just to avoid legal complications or merely follow procedures for the purpose of having a protocol approved by the review committee. To do that would not be compassionate toward the child. The researcher should take the view that no research work is too complicated for a child to understand and, if the researcher is not capable of explaining her basic ideas to a child, then it is likely the researcher does not even fully understand her own work.

As for Buddhism, a typical attitude toward children is not that much different from the attitude of a typical adult toward them. That is, Buddhism does not seem to have much to say in particular about children.

The passages in the Scripture (the *Tipitaka*) that mentions children often do so either as part of a story or as a metaphor for being unexperienced. For example, a passage says that an illustration of impossibility is for children to sail a ship.[4] This shows that the attitude is that children are weaker and less experienced. The typical Buddhist attitude toward all stages of life is that life is full of suffering. The four main stages of life are the epitomes of suffering, namely birth, old age, sickness and death.

Children appear to be closer to birth, thus it can be seen that they suffer too because birth itself is suffering. What is really meant by 'suffering' in Buddhism then needs to be sorted out because it has led many observers to declare that Buddhism is a pessimistic religion bent on seeing everything as full of pain and suffering. Birth is taken to be an epitome of suffering because to come to the world is to experience sorrows, griefs and all other kinds of uneasiness and unsatisfactoriness. Besides, the process of birthing itself is painful and full of suffering because the mother has to endure tremendous pain during childbirth, and the child endures a lot of suffering too, because he or she is being pushed and pulled through the very narrow birth canal. Thus even if the children may be outwardly happy and smiling, they are mired in suffering because their conditions are not stable.

They may be happy at one stage and crying out loud in misery the next. This changeability in life is one condition of suffering mentioned by the Buddha.

All this may sound pessimistic, but the purpose of pointing out that birth is a kind of suffering is to teach the practitioner to see the pointlessness and unsatisfactoriness of life, which will provide motivation for them to continue practicing until eventually they arrive at the ultimate goal, nirvana, which is the state totally beyond any suffering.

So the point for research ethics and bioethics is that Buddhism sees children to be in need of special care and protection because of their vulnerable status. Since they also suffer, they evoke our compassion because children are also connected intimately with other humans and all other sentient beings. They are moreover in need of more active compassion because they are still young and are not able to do things by themselves as adults can.

[4] Atthana-Jataka, about impossible things, from Khuddaka-nikaya, Jataka, Book I. http://84000.org/tipitaka/pitaka_item/v.php?B=27&A=5043&Z=5068. Accessed 6 December 2011.

11.3.2 Women

Unlike children, Buddhism has a tremendous amount to say about women. The Buddhist Scripture, the *Tipitaka*, consists of three main parts, namely the *Vinaya*, the *Sutta* and the *Abhidhamma*.

The *Vinaya* consists of monastic rules for monks and nuns; the *Sutta* consists of the main body of the Buddha's teaching, which was committed to memory by his students during the Buddha's teachings when he was alive. The *Abhidhamma* contains teachings the in form of abstract treatises. What is interesting to us is that much of the *Vinaya*, the monastic rules, contains many rules pertaining women. Since monks must be celibate, there are many restrictions intended to make sure that they stay in line and do not cause trouble because of their attachment to women. One of the most serious offenses for a monk is his violation of the celibacy rule. A monk who has sexual intercourse with a woman will be expelled from the order. Furthermore, there are lesser offenses related to women, such as if a monk touches a woman, then he will be penalized but not expelled, and so on. This shows how seriously the Buddhist monastic order views the woman problem.

In order to maintain order for monks in a monastery which may contains hundreds or thousands of monks, all of whom must be strictly celibate, it is quite understandable how seriously these rules pertaining to women are. Hence in general the Buddhist attitude toward women, at least as appears in the *Vinaya* rules, is that women are the "enemy of the practice," meaning that they represent a threat to the monks because they arouse sexual attachment in the monks, tempting them to stray from their celibate vows. This does not mean that Buddhism denigrates women; it only means that the women represent a threat when one is intent on defeating one's own attachment to worldly desires. If one who is practicing to avoid worldly desires is herself a woman, then men represent a threat to her in the same vein because of men's potential to arouse sexual attachment in the woman too.

Nonetheless, women are also a traditionally vulnerable group because of their still inferior status in many societies as well as their physical constitution. Thus both the UNESCO Declaration and the CIOMS guidelines pay special attention to their situation. The text of the UNESCO Declaration does "[recognize] that an important way to evaluate social realities and achieve equity is to pay attention to the position of women." (UNESCO 2005) This shows clearly that paying special attention to the status and conditions of women in society is necessary for any kind of social equity. The CIOMS Guidelines, however, are much more detailed. According to the Guidelines,

> Investigators, sponsors or ethical review committees should not exclude women of reproductive age from biomedical research. The potential for becoming pregnant during a study should not, in itself, be used as a reason for precluding or limiting participation. However, a thorough discussion of risks to the pregnant woman and to her fetus is a prerequisite for the woman's ability to make a rational decision to enroll in a clinical study. In this discussion, if participation in the research might be hazardous to a fetus or a woman if she becomes pregnant, the sponsors/investigators should guarantee the prospective subject a pregnancy test and access to effective contraceptive methods before the research commences. Where

> such access is not possible, for legal or religious reasons, investigators should not recruit for such possibly hazardous research women who might become pregnant (CIOMS 2002).

The idea behind this guideline is that the woman should be protected because of her special position as one who gets pregnant, which is also a condition that makes her especially vulnerable. A noticeable aspect of the guideline quoted here is that if there's a possibility that the woman who is recruited as a research participant might become pregnant during the course of the research and where the pregnancy might endanger the health of the woman and the fetus, then the woman should not be recruited unless contraceptive methods are available.

In Buddhist countries this is not much of a problem because Buddhism as a whole does not have any qualms against contraception. Preventing birth from happening is not actually an 'unwholesome' (*akusala*) action such that the perpetrator of the action will receive serious consequences afterwards. (However, many Buddhists feel that if conception has taken place it would be wrong to take the life of the embryo, since that is to take the life of a living being, which is against one of the precepts of Buddhism.) Again the idea is that the vulnerable, in this case women, should be protected because doing so would maintain the ideal of equity among all the population groups, as well as of respect for the dignity of all human beings.

11.3.3 The Elderly

Another group sometimes considered vulnerable is the elderly. As mentioned before, Buddhism views old age as another of the life stages which clearly reveals that life is full of suffering and dissatisfaction. In the Buddhist Scripture it is specifically mentioned as one of the undesirable aspects of life that indicate suffering in concrete terms.

Old age is contrasted healthy youth in that the former consists of wrinkled skin, loose teeth, white hair, general decay of the body, and so on, all of which point to the unsatisfactoriness of old age. This is often mentioned in the Buddha's teaching to help people realize that no one will be able to escape these conditions; hence one should start practicing the teaching in order that one understands fully that these are unavoidable natural conditions due to the general nature of the body and of the things in the world. This realization then is among the first stages along the path to eventual Liberation.

The picture of advanced old age painted in Buddhism is intended to show the unsatisfactory nature of old age where one loses one's bodily function and becomes weaker and more frail. However, in all cultures there is also a positive image of the elderly as those who are more experienced and wiser. They are consulted when the community would like to have the benefit of past experiences and the wisdom that is associated with old age. One has an image of an elderly man who tells his children and grandchildren the secrets of the meaning of life. This image shows that being elderly in itself is not a condition of absolute vulnerability. If one is healthy and is able to participate well in community, then one is not vulnerable. It is only when there are such conditions that specifically make them vulnerable that they become so (Walsh 2009).

In the context of research and a clinical setting, being advanced in age *per se* does not seem to be an especially vulnerable position. If the elderly person is healthy enough and is able to live a healthy, active life, then he or she is no different from those less advanced in age who are also healthy and active. However, if the elderly become ill, are institutionalized, or fall prey to social and economic conditions of hardship, then they become vulnerable. The general conditions of old age mentioned in the Buddhist text—that of having wrinkled skin, loose teeth, white hair, and so on—are natural features of old age which by themselves can make someone vulnerable.[5] Loose teeth, for example, can result in the elderly not being able to eat and chew properly, so they would need some kind of care.

Though these conditions could be prevented to a certain extent through modern medical intervention, the fact that the elderly by nature are susceptible to these conditions, and are in need of intervention, seems to show that the elderly could perhaps potentially be vulnerable even though they might outwardly look vigorous and healthy.

According to the CIOMS Guidelines,

> Elderly persons are commonly regarded as vulnerable. With advancing age, people are increasingly likely to acquire attributes that define them as vulnerable. They may, for example, be institutionalized or develop varying degrees of dementia. If and when they acquire such vulnerability-defining attributes, and not before, it is appropriate to consider them vulnerable and to treat them accordingly (CIOMS 2002).

This view still recognizes that if the signs that show that the elderly become weaker and less healthy do not yet appear, as a result of good personal care for example, then the elderly in themselves do not appear to be vulnerable. However, the increased likelihood that the elderly can fall prey to debilitating conditions usually associated with old age shows that they *can* become vulnerable. This is why the last sentence in the passage quoted above says that they should only be considered vulnerable if they show these signs of debilitation.

11.3.4 The Disabled

The last group to be considered here are the disabled. They are clearly vulnerable because of their conditions, which make them less able to function in the same way as normal people.

What come to most people's minds when they think about people with disabilities are images of blind and deaf people, or people who have lost their limbs. How-

[5] There are many places in the Tipitaka where the Buddha speaks of old age. A passage here is typical: "O monks, what are old age and death. Old age, conditions of old age, loose teeth, white hair, wrinkled skin, decay of age, decay of the organs in the animals—these are called old age." Vibhanga-Sutta, from Nidana-varga, Samyutta-Nikaya. http://84000.org/tipitaka/pitaka_item/v.php?B=16&A=33&Z=87. Accessed 6 December 2011.

ever, according to Gregor Wolbring, there is another kind of disability which is social, and not directly physical. In this sense one is also disabled when one is stuck in a socio-economic condition which renders one unable to function properly in society as do those living with "normal" conditions. The underprivileged and the poor, then, are *disabled* because of their being put in the condition which makes them incapable of performing many tasks that should be routine for those who are not in the same condition. The parallel with physical disability is that one who has lost his legs, for example, is unable to walk; in the same vein, one put into the socially disabled condition is unable, for example, to enjoy many of the perks and advantages of life that those living with a better socio-economic condition do routinely enjoy.

For Wolbring the term 'ableism' refers to a set of practices, attitudes, and beliefs that are based on the view that certain forms of ability of bodily function (such as the ability to walk) should be a defining characteristic of a group given special privileges simply due to the possession of such ability. In the same way as racism favors certain ethnic groups over others because of the mistaken belief in the special characteristic of the favored groups, ableism refers to a set of attitudes that favor certain groups of people who possess certain forms of ability, thereby putting those not having the same placed in an unfavorable position (Wolbring 2011). The notion of ableism is introduced in order to raise the awareness of the public regarding the need of the disabled group that they should be accorded the same rights and treatment as that of the "normal" one. Here one can see that being disabled (in both the physical and social sense) is a vulnerable condition.

The Buddhist teaching on compassion is especially relevant to the disabled group However, one should take care not to let the compassion and the active help that arises out of compassionate feelings be confounded with the feeling that one is superior to the disabled, and are helping them because one feels pity to them. This kind of feeling does not eliminate the unfavorable social condition that has led to the claim of ableism from the beginning, and is also contrary to the Buddhist teaching. In Buddhism when one develops compassion and acts in order to help fellow sentient beings out of compassion, one does so with no regard for one's own self. If one performs any action out of egoistic motive, that is, if one does something to help another because one desires some material or immaterial benefits to oneself, then the compassion here is not pure, and can even result in negative karmic consequences.

On the other hand, if one performs compassionate act out of genuine, altruistic desire to help others with no regard for one's own self, then it is a pure act that is genuinely good. Helping others because one feels pity toward them or because one feels some sense of superiority toward them, is not purely compassionate because one still retains some recognition and consciousness of one's own ego.

The CIOMS Guidelines does not have anything specific to say about the disabled as their own distinct group; however Guideline 19 has the following to say:

> Investigators should ensure that research subjects who suffer injury as a result of their participation are entitled to free medical treatment for such injury and to such financial or other assistance as would compensate them equitably for any resultant impairment, disability or handicap In the case of death as a result of their participation, their dependents are entitled to compensation. Subjects must not be asked to waive the right to compensation (CIOMS 2002).

This guideline does not say anything about those who are already disabled before joining the research. However, it says that in case where the disability or injury occurs as a result of the participation in the research, then the participant is entitled to equitable compensation. One can also extrapolate from this that the treatment toward the disabled as a distinct group should be done with equity and justice in mind. And it is the point of this paper that equity and justice should arise out of compassionate feelings according to the Buddhist teaching.

11.4 Conclusion

This paper attempts to delineate the Buddhist viewpoints toward the four main vulnerable groups, namely children, women, elderly and the disabled. The concept of vulnerability is here defined with an eye toward its original meaning of being in a position which makes oneself more liable to attack and to being wounded.

Thus one is vulnerable in the context of bioethics and biomedical research when one is in a position which makes oneself more liable to attack by diseases or other physical threats, or by various forms of social determinants such as being poor and disempowered. Then we discussed how Buddhism looks at the four main groups. The key idea in understanding how Buddhism views these vulnerable groups is that of compassion, the feeling that one should help others and care for them, relieving them of suffering, is based on the understanding that everything is interconnected and is in the very real sense one and the same. The understanding of interconnectedness or interdependence leads naturally to the feeling that one does not stand alone in the world, but in fact what we normally take to be our own selves are but constructions that are contingent and always liable to change, and our putative selves here can remain only through relationships with and kindness from others. Thus it is necessary that we pay the kindness back to all other beings in return.

Acknowledgement I would like to thank the organizers of the UNESCO Workshop on Human Vulnerability for their generosity in inviting me to the Workshop, as well as the partial support given me by Chulalongkorn University, through a grant in the National Research University Project, grant no. AS569A.

References

CIOMS. 2002. International ethical guidelines for biomedical research involving human subjects. Bull Med Ethics 182:17–23. http://www.recerca.uab.es/ceeah/docs/CIOMS.pdf. Accessed 31 Aug 2011.

Coleman, D. L. 2007. The legal ethics of pediatric research. *Duke Law Journal* 57 (3): 517–624.

Kahn, J. P., A. C. Mastroiani, and J. Sugarman. 1998. *Beyond consent: Seeking justice in research*. New York: Oxford University Press.

Macklin, R. 2003. Bioethics, vulnerability and protection. *Bioethics* 17 (5–6): 472–486.

Solbakk, J. H. 2011. Vulnerability: A futile or useful principle in healthcare ethics. In *The SAGE handbook of healthcare ethics*, eds. R. Chadwick, H. ten Have, and E. M. Meslin, 228–238. London: Sage.

UNESCO. 2005. Universal declaration of bioethics and human rights. http://portal.unesco.org/en/ev.php-URL_ID=31058 & URL_DO=DO_TOPIC & URL_SECTION=201.html. Accessed 31 Aug 2011.

Walsh, S. A. 2009. Conducting research with the elderly: Ethical concerns for a vulnerable population. *Southern Online Journal of Nursing Research* 9 (4): 1–13. http://www.resourcenter.net/images/snrs/files/sojnr_articles2/Vol09Num04Art03.pdf. Accessed 2 Sept 2011.

Wolbring, G. 2011. Ableism and ability glossary. http://ableism.wordpress.com/ableism-glossary. Accessed 2 Sept 2011.

Chapter 12
A Christian Consideration of Human Vulnerability in Healthcare and Research

Colleen M. Gallagher

12.1 Introduction

The International Bioethics Committee (IBC) called for continuing recognition of human vulnerability in healthcare and bioethics sciences in the recent report of June 2011 (UNESCO 2011). The report yielded special attention to factors that determine special vulnerability, access to healthcare, the provision of appropriate healthcare, inequalities of power, and vulnerabilities in human research. All of which are true in the use of technologies used to treat illness. Although the IBC is not a religious organization following a specific set of moral teachings or tradition, the following will consider IBC principles through Christian perspective and its scripture as illustrative.

Vulnerability is often greater when the individual or population involved loses abilities due to illness or other factors that take away from the usual means of maintaining and growing in all manner of being. It can be defined as "(1) the capability or susceptibility to being wounded or hurt, as by a weapon, (2) being open to moral attack, criticism, temptation, (3) open to assault; difficult to defend" (Morehead 1995). Some people are more vulnerable to conflict than others; this vulnerability arises from a lack of understanding their role as an individual and as a member of society. By paying meager attention to resolving their own interpersonal conflicts, they fail to realize the broader community implications of individual discord (Dobson 1992).

The Catholic-Christian tradition writes many of its teachings addressed to its leaders and to "all people of good will," as the teachings reach far beyond the Catholic Church's members. One of the recent teaching documents is *Evangelium Vitae*. In the first chapter Pope John Paul II discusses the contradiction of global documents calling for social, cultural and political change toward a culture that promotes life and human dignity, and the practices which prevent such a culture.

C. M. Gallagher (✉)
Section of Integrated Ethics in Cancer Care,
The University of Texas MD Anderson Cancer Center, Houston, Texas, USA
e-mail: CMGallagher@mdanderson.org

J. Tham et al. (eds.), *Religious Perspectives on Human Vulnerability in Bioethics*,
Advancing Global Bioethics 2, DOI 10.1007/978-94-017-8736-9_12,

> In our service of charity, we must be inspired and distinguished by a specific attitude: we must care for the other as a person for whom God has made us responsible. As disciples of Jesus, we are called to become neighbors to everyone (Luke 10:29–37), and to show special favour to those who are poorest, most alone and most in need. In helping the hungry, the thirsty, the foreigner, the naked, the sick, the imprisoned – as well as the child in the womb and the old person who is suffering or near death – we have the opportunity to serve Jesus. He himself said: "As you did it to one of the least of these my brethren, you did it to me" (Matt 25:40). Hence we cannot but feel called to account and judged by the ever relevant words of Saint John Chrysostom: "Do you wish to honour the body of Christ? Do not neglect it when you find it naked. Do not do it homage here in the church with silk fabrics only to neglect it outside where it suffers cold and nakedness" (John Paul II 1995).

The encyclical uses generations of Christian theology and human lived experience to discuss the many ways that human life has been devalued and destroyed. It specifies the many ways the human community is vulnerable at this time in history. The pope calls upon all in the human family to recognize these challenges and gives several actions that can be taken by individual Christians as, well as by corporations and governments, to alleviate human suffering and bring society as a whole to address human vulnerability. Similarly, in 2005 Pope Benedict the XVI stated:

> There are many forms of voluntary assistance, models of mutual service, of which our society has urgent need. We must not, for example, abandon the elderly to their solitude, we must not pass by when we meet people who are suffering. If we think and live according to our communion with Christ, then our eyes will be opened. Then we will no longer be content to scrape a living just for ourselves, but we will see where and how we are needed. Living and acting thus, we will soon realize that it is much better to be useful and at the disposal of others than to be concerned only with the comforts that are offered to us. I know that you as young people have great aspirations that you want to pledge yourselves to build a better world. Let others see this, let the world see it, since this is exactly the witness that the world expects from the disciples of Jesus Christ; in this way, and through your love above all, the world will be able to discover the star that we follow as believers. (Benedict XVI 2005).

The application of Christian scripture is important due to its shared attributes in teaching Christian traditions. There are many stories of Jesus' healing the vulnerable and teaching all about how we can better care and provide for one another maintaining equality of health, spirit and society. In Matt. 25:31, the story of the last judgment, hunger, thirst, nakedness, being a stranger and illness are each mentioned. These are all situations in which humans are vulnerable. Each is a time of additional powerlessness. Children are also vulnerable as they too do not have equal power to adults who make decisions and are charged with caring for the young (Matt 19:13–15).

12.2 Access to Healthcare

The nature of vulnerability involves a shortened lifespan, poor quality of life and the risk of imminent death according to the IBC (UNESCO 2011). Causes cited include the inability or failure of states to provide adequate supplies of drugs and delivery of education regarding known prevention strategies. The experience of vulnerability creates stress and anxiety which affects physiological, psychological and social functioning. It is affected by personal factors as well as factors within the environment. Trends in society indicate that increasing numbers of vulnerable people will create additional demands on an already over-burdened health care system. A shortage in resources can lead to further abated access to healthcare (Rogers 1996).

Access may involve mobilizing people to the needed treatment resources or bringing the resources to the people. The resources can be present, but difficulties exist in connecting these resources with the ill. In the Gospel of Mark, a paralytic man in the city of Capernaum could not get to Jesus to be healed as the throng of people seeking assistance was so large that doorways were blocked. The man required four friends to help him go from place to place as he could not walk. The friends went atop the roof and found a place where they could remove thatch. They made a hole and lowered their ill friend on a mat so that he was put before Jesus for healing (Mark 2:1–12). Access to healthcare is not something that an individual can necessarily navigate alone. It involves a community of people to make the changes needed.

Nearly all the biblical stories about physical healing involve at least one person asking Jesus to heal another. Healthcare is a communal experience and responsibility. At the same time, it is also involves the individual's responsibility. Mark also tells the story of a woman with a hemorrhage who approached Jesus on her own, just wanting to touch his cloak and knowing she would be healed (Mark 5:25–34). In this story one can note that it takes great faith to have miracles or that the woman, after seeking many types of help, continued to seek assistance. She was continually open to the possibilities even when she was exhausting other methods and traveling distances for her care. What is common in both situations is the care involved a relationship between the ill person, the community and the healer.

A community is needed to help advocate for individuals. In the United States there have been several recent arguments surrounding health care for immigrants and for those who are currently underserved. A communitarian approach has been called for by several social and religious leaders. Reflections related to several of the issues including political action that addresses utilization of resources, the ethics of medicine, and community's ability need to be part of the process, as well as a communitarian perspective. (Kuczewski 2011)

12.3 Disadvantaged People in Developed Countries

When considering disadvantaged people in developed countries the IBC report describes remedies for the vulnerability of unavailability of healthcare resources because of a lack of ability to pay and people not receiving full information about their care options. This disparity arises amongst the vulnerable population.

The word "disparity" can be defined as "the condition or fact of being unequal, as in age, rank, degree, or religion." While disparity in health care has been closely associated with equity, there are several potential reasons for the differences observed at the individual level. Factors such as race, ethnicity, income, education, place of residence, age, religion and others affect quality of and access to health care are both independent and additive (National Healthcare Disparities Report 2010).

In the Gospel of Luke the use of money (in this case, the resource) to benefit self and others is discussed (Luke 19:11–27). A wealthy man was going away from home for a long period of time. He gave some of his servants money instructing them to invest the funds. Some took the money for themselves, some invested and doubled the money, others invested and made a little less. All those who invested and worked to have the money make more money for the rich man were rewarded. One servant hid the money and did not invest it. He wanted to maintain it without risk. This man was harshly dealt with. The moral of the story was proclaimed that "whoever has will be given more, but the one who has not will lose the little he has…" The developed world, having resources, is called to share what it has to provide appropriate diagnostic and treatment resources.

The information about alternative and therapeutic opportunities is another aspect of power and powerlessness. The healthcare community has knowledge of these and it is perceived that others do not. The situation of patients not always receiving full information remains a challenge due to many reasons, not solely attributed to lack of funding. Healthcare providers are called to their professions and set out to do specific work. The specialties of healthcare actually may cause a narrower view of medicine, and prevent a provider from knowing all the opportunities. Thus the responsibility is to be shared. Once again, the Christian scriptures describe the calling forth of those who would be leaders and sending them out to bring healing of different kinds to the people. Luke tells the story of the seventy-two where the disciples are reminded that they are to take with them only that which they need, to eat what is shared as their only payment. Jesus gives to them the authority to heal body, mind and spirit. They are reminded to remain humble so that the power and authority given them is not abused (Luke 10:1–20). The story is repeated in other gospels as well.

12.3.1 Migrants

Migrant people are especially vulnerable according to the IBC. One of the key questions regards whether they have a right to be treated and to what extent their treatment will be paid. Their marginalization in society is due to many reasons which includes but is not limited to language, knowledge of culture, education levels, and economics.

Fuller-Thompson reported a 15 % two-step decline in health among immigrants during the first four years after arrival in Canada compared to the 6 % similar decrease in health for longer term immigrants. Asgary and colleagues found similar results when considering immigrants from West Africa and Central America to New York City (Fuller-Thompson 2011).

In the Christian scriptures we find several stories in which Jesus befriends an outcast, stranger, or person who has been marginalized because of their heritage. One of those stories is the story of the Samaritan Woman (John 4:4–42). In this story Jesus stops at a well to rest while on one of his journeys. Jews and Samaritans were to have nothing to do with each other. Jesus spoke to a Samaritan woman asking for water.

They had conversation while no Jews were around the well. When the friends and followers of Jesus returned to the well they were surprised and questioned Him about why He spoke with the woman. The woman went into the town and told people of this man who spoke to her and that he was a special person. If we look at the story in relation to migrant people and their marginalized status in our world today, we can consider that when one is in need it does not matter if they are otherwise marginalized. We are to assist when able. This recognizes the marginalized person maintaining, and sometimes reminding them, of their integrity. It also gives spiritual depth to those who render aid and support.

12.4 Unfair Allocation of Resources

The unfair allocation of resources can allow people to succumb to illness more quickly than if healthcare and health education was made more available. The IBC notes that "States should have in place a robust resourced healthcare system that fairly and without discrimination provides adequate care to all citizens."

This is a very large challenge no matter the development of the countries. There are many purposes that require funding and not all people recognize the same importance for each of the purposes. The statement also appears to be in contrast to the call to address healthcare needs of the migrant people as cited above, as it limits itself to citizens. However, Christian scripture does contain a story from the life of Jesus in which He entered the temple and is angered by the buying and selling of goods and animals for sacrifice that did not match the actual purpose of the temple. He was so angered that He talked about the destruction of the temple (John 2:13–23).

In today's world we often hear politicians and business leaders discuss the destruction of the existing systems so that new, more just systems can be structured.

Inequality of Power Distinctions in power are cited and discussed by the IBC as disrespect for the patient's will and professional self-interest. These are not necessarily in opposition, but often are.

A patient who needs treatment for a particular disease that has no known cure may enter into research protocols that are of personal interest to the healthcare provider as a professional goal. However, when the medical professional or company holds the power to grant or not grant treatment it is essential to pay strict attention to the need for patient autonomy related to their involvement in the study. Mark tells the story of the wealthy man who wants to know what to do to get into heaven (Mark 10:17–25). The man proclaims that he has done the required things listed in the Hebrew scripture known as The Ten Commandments. Jesus tells him there is something more he must do, sell what you have and give to the poor, then follow Jesus.

Basically, those who enter into healthcare professions in any way are called to more than minimum standards. They are called by their professions to have integrity, to live their work recognizing the integrity of the persons they care for as well as their own. If this is followed the power difference is diminished and the self-interest also decreases, to the benefit of the patient.

12.5 Today's Bioethical Research Challenges

By looking at some of the questions in bioethics today, especially in research, the inequality found can be greater than in healthcare that does not involve research. One of the most challenging research questions of the day is the use of stem cells to treat illness and/or to enhance the human expression of strength. The stem cells in question are those that are pluripotent, those that can grow to become different types of cells in the body.

The use of stem cells from persons who are already born, or the cells from the umbilical cord of a born child is questioned related to purpose. For the treatment of most diseases there is not much question. It is when the purpose is to genetically enhance the human being's abilities that the questions arise. The integrity of the human person physically and mentally is itself questioned. Stem cells from human embryos are questioned because it involves the destruction of a human being when it is totally vulnerable both physically and by the decisions of others as it has no voice of its own. If we follow the thinking of the principles of human vulnerability and personal integrity as voiced by most Christians the most vulnerable are to be protected. The Gospel of Mark tells us that Jesus blessed the children noting that they are not to be hindered and that Jesus embraced them and placed His hands on them (Mark 10:13–16). Jesus cured the twelve year old child of Jairus (Luke 8:40–42) as well as other children.

Neurotechnologies are growing in their use. One of the key questions in their use for treatment as well as for research is how it changes to mind and/or personality of the person. Thus is it essential that attention be given to the autonomy of the patient as well as to preparation for their family members and friends as the person and personality of the patient may be changed beyond recognition. Experimentation must be meticulous and longitudinal in nature and be done on as few individuals as possible to still meet the scientific goals.

Genetic research involving the change of DNA in a patient is quite different than genetic research which determines what the genetic reason is for disease and how to treat those diseases. At the moment the use of genetic information requires protection of the information if the information will negatively affect the patient socially and sometimes psychologically. The larger challenge of today is the request of companies who want the tissues for future research without explanation to the donor/patient and that have no benefit for them. This may continue the lack of balance of power between patient and those who choose whether treatment will be given. Many times the companies have information that the patient's doctor does not have. This violates the patient-physician relationship The healing elements of the relationship are actually removed.

There are always unexpected risks in human research. Responsibility for known and unexpected risks needs to be delineated prior to the research taking place. Even in research there is a relationship, although it is often with an unknown healthcare provider or a company.

We as humans live in societies. These are as small as families and as large as nation-states. We are human and as such owe to each other dignity, integrity, and respect.

12.6 Conclusion

Special vulnerabilities are recognized by most people, states and religious traditions. The challenge of the guiding principles of human vulnerability and personal integrity is often one of knowing whether these are characteristics of persons or the condition of being human. Scientists, healthcare providers and some traditions consider them characteristics of being human and as such are challenged in the protection of humans. Most religious traditions, including Christianity, consider the vulnerability and honoring of integrity as conditions of being human and therefore are often at odds with scientific movement when it does not give protections to humans as vulnerable and having internal integrity.

References

Pope Benedict, XVI. 2005. Libreria Editrice Vaticana. Apostolic journey to cologne on the occasion of the XX World Youth Day: Let us go forward with Christ. Homily at closing mass of World Youth Day. Last modified August 25, 2005. http://www.vatican.va/holy_father/benedict_xvi/homilies/2005/documents/hf_ben-xvi_hom_20050821_20th-world-youth-day_en.html. Accessed 6 March 2012.

Dobson, E. 1992. *Mastering conflict and controversy*. Portland: Multnomah.

Fuller-Thompson, E. 2011. Health decline among recent immigrants to Canada: Findings from a nationally-representative longitudinal survey. *Canadian Journal of Public Health* 102 (4): 273–280.

John Paul, II. 1995. Evangelium vitae. http://www.vatican.va/holy_father/john_paul_ii/encyclicals/documents/hf_jp-ii_enc_25031995_evangelium-vitae_en.html. Accessed 25 March1999.

Kuczewski, M. 2011. Who is my neighbor? A communitarian analysis of access to health care for immigrants. *Theoretical Mediine and Bioethics* 32:327–336.

Morehead, A. 1995. *The new American webster handy dictionary*. 3rd ed. New York: Penguin Putnam Inc.

National Healthcare Disparities Report. 2010. *Access to healthcare*. Washington, DC: Government Printing Office. (November 2000).

Rogers, A. 1996. Vulnerability, health and health care. *Journal of Advanced Nursing* 26:65–72. (New Brunswick).

United Nations Educational, Scientific and Cultural Organization (UNESCO). 2001. Report of IBC of the principle of respect for human vulnerability and personal integrity. International Bioethics Committee (IBC), pp 1–13.

Chapter 13
Christian Perspective on Vulnerable Groups: The Elderly and the Disabled

Hans Ucko

13.1 Introduction

Elderly people may be handicapped but new-born babies may also be handicapped. We are all aging, depending on how long we live, but we are not all necessarily facing a handicap.

What are elderly and handicapped saying about having as a major conjunction between themselves that they are vulnerable? Surely, elderly or handicapped are much more than vulnerable. I ask myself whether this is not a question of reducing the identities of both elderly and handicapped. From other contexts I know that reducing identity to one singular attribute is fraught with difficulty. There is a risk when we are being reduced, or reduce ourselves, to one identity. When reading various social or political contexts in society and in the world, we see that elevating one's identity as a banner, often entails an Other to be the very opposite of what one is oneself.

We are in our time living through many vehement assertions of identity. A particular politics of identity based on a sense of victimization, reducing identity to a single affiliation, facilitates the creation of "identities that kill", says Amin Maalouf (Maalouf 2000). We see it often. The resentment of the West in many parts of the Arab world and vice versa is an obvious example. Migrants and *Gastarbeiter,* even those for several generations in Europe, feel marginalised, when being reduced to this one particular identity. It furthers a self-image and identity, which is only defined as underdog or victim.

When my self-image is one of a victim, the whole field of vision is narrowed down and the horizon is lost. One does not find one's way out. "The most powerful weapon in the hands of the oppressor is the mind of the oppressed", as Steve Bantu Biko rightly said it. (Biko 2002).

Taking with extreme care this question as our topic, I ask myself whether bringing together elderly and handicapped as one vulnerable group is not creating

H. Ucko (✉)
Geneva, Switzerland
e-mail: hansucko@gmail.com

J. Tham et al. (eds.), *Religious Perspectives on Human Vulnerability in Bioethics,* Advancing Global Bioethics 2, DOI 10.1007/978-94-017-8736-9_13, feelings of alienation, that that which is one characteristic or attribute is taken to be the only one, identifying me, saying who I am, *pars pro toto*. Reduced to one identity, one of a victim, whether elderly, because society favours youth, or handicapped, because society favours people without handicap. Would it be appropriate to ask whether vulnerability is enough to characterise both elderly and handicapped? And are elderly only elderly, and handicapped only handicapped?

13.2 The Elderly as a Vulnerable Group

The world's population aged 60 or older is expected to double by 2050, to a record of 21 %—representing 2 billion people. Certain topics appear consistently around the world in relation to elderly: discrimination, right to health, violence and abuse, social protection, age and social exclusion. There are concerns about the status of older persons revolving around their social and economic situation, their participation in development and their relationship to human rights.

Concerns have been raised about age discrimination and neglect and, in response, the empowerment of older persons is already, and will become even more a central, dimension in constructing their human rights and their participation in development. Not only because elderly are visibly more present in society than before, but the aim needs to evolve towards perceiving older persons as active members of society, and not just as recipients of charity and welfare.

This is important since, at the age of 80, women outnumber men by two to one, and are the most vulnerable, facing more social stereotypes, participating less in society and benefiting from fewer available opportunities. There is a prevailing culture aimed at youth, and older people have gone through a progressive exclusion and marginalization. Elderly people feel that they become "a burden." They feel useless. Age is looked upon as an incurable disease. This has led us into a situation, where we fumble for words and are unsure of how to describe and identify the particular characteristic of the human rights of older people.

Another important gap to be covered with respect to rights for older persons refers to their social security coverage. Elderly people often fall prey to chronic diseases that, more often than not, are painful and even performing routine tasks becomes difficult. As our society becomes more "grey," who will pay the increasing medical bills for the elderly? Longevity is increasing the gap between retirement and death, and Social Security may not be able to pay. Can we provide coverage for more and more elderly people with fewer and fewer younger people providing the means to cater to the elderly?

Acknowledging that the way of life is that elderly people become more dependent, we still live in times where being dependent is looked upon as a sign of weakness, and many are the elderly who resent that this day is coming, when there is no other choice. And the question arises whether there are health care options that can preserve some semblance of independence while, at the same time, treating their medical conditions.

Religious scriptures and traditions address aging; it is part of religiosity itself; religion attempts to address what it means to be a human being, where we come from, what our life is all about, where we are going. It is in this context that the question of the elderly is asked and answered. What does it mean to grow older? What are the insights and ways of elderly people? Attitudes toward aging in general, and toward elderly people in particular, are present in all religions. Religion, perhaps because more elderly people rather than young people cultivate it, looks upon elderly people not as a problem to overcome but as a resource for a true life. Scriptures hold dimensions that lift up age as a sign of grace: age is beautiful, age demands respect and dignity, elderly people are giants of the forest, they are wise, full of experience, worthy of our praise and admiration.

The Bible teaches that old age is the blessing of God. In Proverbs 23:22, Solomon exhorts his son to "harken to your Father who begot you and do not despise your mother when she is old." In the story of Job, we find that Elihu, the younger of Job's friends, waited until the older men had spoken to Job and, since Job was his elder, he also tempered his communication to Job with admiration and respect. Old age may be the fruition of a moral life and an indication of God's favour. "Follow the whole instruction the Lord your God has commanded you, so that you may live, prosper, and have a long life in the land you will possess" (Deuteronomy 5:33).

Old age is a general part of God's purpose for a normal life. "He took his last breath and died at a ripe old age, old and contented, and he was gathered to his people" (Genesis 25:8). "David, son of Jesse…died at a good old age, full of days, riches, and honour…" (1 Chronicles 29:26, 28). "Then Job died, old and full of days" (Job 42:17). The so called Isaiah vision (Isaiah 65) looks forward to the time, when "no more shall there be in it an infant that lives but a few days, or an old person who does not live out a lifetime; for one who dies at a hundred years will be considered a youth, and one who falls short of a hundred will be considered accursed" (Isaiah 65:20).

In Exodus 20:12 we find the commandment, "to honour your father and mother, that your days may be long in the land which the Lord your God gives you." In Mark 7:10, we find Jesus saying, "Honour your father and your mother"; and, "Whoever speaks evil of father or mother must surely die."

Old age should be a time for ego integrity. The Hindu tradition encourages the elderly at the fourth stage, at the fourth ashrama, to become and embark upon the "sannyasa" or the wandering ascetic stage. The elderly have made their mark on life. They have performed well. They have confidence that their life was well spent in raising children, that they made the world a better place, and that they have enabled the next generation to continue. Elderly have a wealth of wisdom to share, experience to relate, expressions and advice on life to share. They await new experiences, and are getting prepared for the last experience of this life. They have earned love and respect.

Nature herself teaches us that age demands dignity and honour. The silverback is the one to respect and honour. The older the tree is, the more majestic it is. Old wines and cheeses are praised and honoured for taste. Society is losing out when embarking upon a cult of youth, neglecting old age and even more not facilitating any interaction between the young and old generation, when young people do not get to know the elderly.

Age does not mean that someone is useless. Young should be encouraged to benefit from the talents of the elderly. We should allow elderly people to share their life experiences. We should remember the wisdom of St. Paul to Timothy: "Do not speak harshly to an older man, but speak to him as to a father (and) to older women as mothers" (1 Tim.5: 1–2) and how he emphasizes the same in saying about one of the commandments in the Decalogue: "Honour your father and mother"—which is the first commandment with a promise—"that it may go well with you and that you may enjoy long life on the earth" (Eph. 6:2–3). The Bible thus proclaims that the elderly will receive the respect, dignity and honour they deserve, that someone will care and want to return the love and care that they gave to others.

The Bible is at the same time consistent and clear in its message about those who are least able to fend for themselves. Elderly people as a group would fit into this category. It is true that in recent years, various lobbying groups have served both to increase elderly people's visibility and to provide much-needed political clout. Nonetheless, decreasing physical vigour and steady or declining income mean that elderly will likely continue to be a vulnerable population. But whatever response we can give to this fact, one thing is certain: the way ahead is to begin cultivating consciously in society the interaction across generations.

With this I turn to the second part of my assignment: the handicapped as vulnerable groups.

13.3 The Handicapped as Vulnerable Groups

I would like to begin with the word 'handicapped' itself. At one point in the ecumenical movement the very word handicapped came to be questioned and the word disabled was recognised to be a more appropriate term to describe the condition of "handicapped groups."

Building upon the "International classification of impairments, disabilities, and handicaps: a manual of classification relating to the consequences of disease" (Geneva; World Health Organization 1980), the attempt to define the categories could lead to saying that a *disability* is an inability to execute some class of movements, or to pick up sensory information of some sort, or perform some cognitive function, that typical unimpaired humans are able to execute or pick up or perform, and that a *handicap* is an inability to accomplish something one might want to do, that most others around one are able to accomplish.

The term "handicap" is sometimes now avoided and its use is conditioned on who uses it and how it is used. One can be handicapped, but not be disabled. And one can be disabled, without being handicapped, if the proper tools and supporting structures are provided. The realities of handicap and disability have a much wider reach than we usually think. Thanks to glasses and hearing aids we can reduce some of the impact of handicap and disability.

In the ecumenical discussion, there was also an attempt to use instead of the term 'handicapped', the words 'differently abled'. This was a term, which probably had good intentions. One wanted to find kinder words rather than using words that sounded denigrating. A handicapped person was seen to be as able as anyone else but in a different way, trying to find words that just point to difference as we can be different in height, skin, colour of eyes, etc. When we talk about handicapped, we who are not handicapped, think that the handicapped are in a situation that is undesirable and adverse and, since this makes those of us who are not handicapped a bit uneasy in our relationship to handicapped, we look for words that so to say sound as inoffensive as possible, words that do not say it as it is but provide protection from reality and mostly protection for ourselves not to have to face the plight of the handicapped.

The stand-up comedian George Carlin once coined the word "handicapable"; it sounded almost nice to be "handicapable." The point of his argument was however to drive home the real truth: you do not change the condition by giving it another name.

The word disability is a modern word and has no direct equivalent in ancient languages. There is not one word either in Hebrew or Greek that means disabilities in the way we use that word today. In the Bible we read about the blind, the deaf, the paralysed: all the injuries that today we gather under the term disability, but the collective term is lacking. Considering the absence of the term disability in the Bible makes it a bit more complicated. Disability is a modern term, built on a modern understanding of people with disabilities as a group of people with certain needs due to lack of different physical or mental functions.

The terms disability and handicap relate to politics which we started to use when modern social welfare policies began to take form, and the need to name a specific group which demanded special action from governments was felt. Even though all the conditions today we term disabilities were well known in both the Old and New Testaments, the persons who lived with these conditions were not understood, either by themselves or by others, as persons with disabilities in the modern meaning of that word. According to our interpretation of the Bible, every human being is created in God's likeness regardless of his or her physical or mental capacities. The love that God has for every being also applies to people with different kinds of disabilities. God's wish to save every human being and the whole of creation from evil is a wish God also has for those who live with disabilities.

This means that God's command that we love our neighbour as ourselves (Lev 19:18) also includes those who have disabilities. According to the same chapter, God commands the Hebrews not to curse the deaf or put a stumbling block before the blind. The Old Testament tells its readers to treat people with disabilities justly and fairly. Moses tells people that to mislead a blind man on a road will lead to a curse (Deut 27:18).

But the Bible makes statements in both the Hebrew Scriptures and the New Testament, and tells stories, that to a modern reader must seem difficult. As with many other issues, the Bible is not everywhere and always unequivocally advocating views we hold today on disability. The Bible as well as other religious scriptures have condoned slavery and conserved the position that women are secondary to men. Building only upon a literal reading of the Bible will not help us look upon disabled people in their own right. The Bible does make statements about disabled people that we wish were not there. This is not easy to handle, either we do not want to recognise such words and stories but brush them away, or we engage in an *apologia pro vita sua,* that is, we are so keen to defend the Bible that we become dishonest.

There is a well-known scripture passage that from our modern point of view will look discriminating, namely the command in Leviticus 21 that no one who has a blemish may be a priest. There are ways to explain away the lack of sensitivity in these words and there are probably ways of reducing the offense but sometimes we should, as religious people, be bold enough to call a spade a spade. We might want to consider whether we would not do disabled people more of a service, if we acknowledged that the Holy Scriptures could not alone or literally be our only moral and ethical teacher.

This is even more important when considering that there are Christians who cling to such passages in the Bible to credit their own world-view. Such passages have been and are still interpreted in such a way that people with disabilities would be unworthy and their injuries or sicknesses a punishment for sin. We can see this interpretation behind the disciples' questions to Jesus when they meet the man who was blind from birth: (John 9:2) "Who sinned, this man or his parents that he was born blind?" Jesus' answer is clear: "It was not that this man sinned or his parents."

Another passage seems to allow for such an interpretation. It is the story in Luke 5 of the paralysed man who was brought to Jesus. The first thing Jesus says to the man when he meets him is "Man, your sins are forgiven you" (Luke 5:20). We need to face such texts openly and not hide them or whitewash them. This is difficult but essential, if we want that people with disabilities should not question the passages, where Jesus heals persons with different kind of disabilities. They ask if Jesus' only interest in persons with disabilities is to heal them. Is the Bible only a book about miraculous recoveries whereas our experiences—the experiences of those whose lives are not success stories—are excluded?

Maybe one answer to the question as to what we can learn about disabilities in the Bible is to see that the question of who is weak and who is strong is much more complicated than the words themselves tell us, and that we need the contribution of all to come closer to a meaning that does not denigrate, but affirms, the dignity and humanity of all. It seems to me that we can in the Bible read an overarching message that Christ came to tear down the walls (Eph. 2:14).

Whenever we consider the ways in which to respond to issues of disability, we would do well to remember the walls that we have set up. All of these walls are human and they contradict Christ's ministry of reconciliation; walls that shut people in or shut people out; walls that prevent people from meeting and talking to others. In days gone by, people with disabilities were actually kept behind walls, inside institutions. And even if, in many societies, the walls have come down, persons with disabilities may still find themselves isolated. There are walls of shame; walls of prejudice; walls of hatred; walls of competition; walls of fear; walls of ignorance;

walls of theological prejudice and cultural misunderstanding. The disabled did not erect these walls; those who wanted to stash away the disabled have erected these walls.

But the tragedy is made even worse when the disabled internalise what those who were not disabled have said about the disabled. It is here that the words of Biko merit being quoted again: "The most powerful weapon in the hands of the oppressor is the mind of the oppressed" (Biko 2002). Dalits for centuries denigrated by the caste structure finished by internalising what was said about those outside the caste-system, and acted as if they actually were inferior.

I will now for the remainder of my presentation refer to a text from the ecumenical movement, "A Church of All and for All—An interim statement" (World Council of Churches Central Committee 2003), which has been written by disabled people, parents and others who experience life alongside them in various ways.

The importance of this statement is not only that it does address the theme of this workshop but also that it does so in challenging the way churches treat "persons with handicaps" and "the differently-abled" and "persons with a disability" (all those terms were designed to reflect inclusiveness and each replaced the other) as *objects* rather than subjects of reflection." It makes sure that "persons with disabilities are themselves the *subjects* or actors of reflection or action."

"*The disabled*" have struggled hard to become recognised as "*disabled people*." Throughout history, disabled people have been de-personalised and perceived as a problem to be dealt with. They are often seen as a homogeneous group whose individual differences do not need to be respected. In common with all groups in society, disabled are also very diverse and have different stories.

Persons with disabilities often become vulnerable to discriminatory social trends. A market economy encourages in many countries systematic abortion of the foetus with certain malformations and those with Down's syndrome, which gives a negative message of society's views of disability.

Disabled people become vulnerable to easy commercial fixes and religious groups which offer miraculous healing in the setting of superficial acceptance and friendship.

In our wrestling with God, as disabled people we all ask the same basic questions, but the theological enquiry involved may be complex. Why my loved one, why me? Is there a purpose to my disability?

We have wrestled with God intellectually and physically to achieve peace, and whilst some have been privileged to write intellectually about it, others have shown it in their innate gift of grace revealed in love and affection toward those who care for them so deeply. If so many disabled people have this ability to come to terms with God, the Church must surely find ways of accepting the gifts, which disabled people have to offer. It is not a case of meeting halfway but of full acceptance.

How can we interpret from a theological perspective the fact that some people live with disabilities? What does that fact tell us about human life in God's world?

Disability has historically been interpreted as loss, an example of the tragedies that human beings can experience. The Gospel stories about how Jesus heals persons with different diseases and disabilities are traditionally interpreted as illustrating how human beings are liberated and empowered to live a richer life.

People with disabilities are seen as weak and needing care. As a result, they are viewed as objects for charity, those who receive what other persons give. Thus, people with disabilities cannot meet other people in the churches on equal terms. They are regarded as somehow less than fully human.

The church has justified this view from different theological perspectives. For instance, disability has been interpreted as a punishment for sins, either committed by the persons with disabilities themselves or by their relatives in earlier generations. Or disability has been understood as a sign of lack of faith that prevents God from performing a healing miracle. Or disability has been understood as a sign of demonic activity, in which case exorcism is needed to overcome the disability. Such interpretations have led to the oppression of people with disabilities in the churches.

To live with a disability is to live with *other* abilities and limitations that others do not have. All human beings live with limitations. Is not disability something that God has created in order to build a plural, and richer, world? Is not disability a gift from God rather than a limiting condition with which some persons have to live? Such questions need to be taken seriously when we are searching for new theological understandings of disability.

What does it mean to talk about the image of God in relation to persons with disabilities? If the image is described as "perfect body", or "perfect reason", how can persons with disabilities embrace such an image of God? What is the relationship between our theological language and practice with regards to the issue of disability? How much of the medical and social language which treats persons with disabilities as objects determines both academic theologies and general attitudes about, and towards, persons with disability as objects of pity, forgiveness and healing? What does it mean to call the Church the body of Christ? Can persons who are visually impaired or who have a body with cerebral palsy be included? Although many Christians consciously deny any relationship between disability and sin (which also includes suffering), some of their attitudes seem to reflect such a link.

Persons with some form of mental disability or some form of learning disability will be disqualified as human beings because they will not reflect the definition of the image of God as soul, as reason or as rationality. A hermeneutic of suspicion cannot accept the image of God or soul as reason or rationality.

In relation to disability, theology is challenged to talk about God, faith and life in a way open to God's future; that can surprise us all, and unite and transcend every human existence. A theological understanding of disability has to interpret this issue in the context of the *unfinished* history of God's salvation.

In our liturgies celebrating the Eucharist, the words of Christ, *Take, eat; this is my body, which is broken for you.* (Cf. Mt 26:26), resonate with holiness and reverence. They recall the theological significance of the Christ-event but they reveal at the same time aspects of what life is also about: brokenness and frailty.

Such language, such metaphors are the very opposite of the temptation to triumphalism, which so often, in different ways and with different impacts, has bedevilled the church, be it in relation to people of other faiths, people of different sexual orientation, women, people of colour, and also through marginalisation and pity: disabled people. Does this mean that we should look upon the condition of being disabled as a given, as a constant? In many cases, disability is exactly that. But it does not rule out that we one day through scientific progress will find ways to discover a cure for handicaps. Meanwhile, the main issue today is learning that difference is not to be looked down upon and frailty is not necessarily to be pitied.

The document 'A Church of All and for All' concludes with the following spiritual reflections:

When we think of people with disabilities, too often we tend to think of people who are weak and require our care. Yet, in his epistles, St Paul implies that weakness is not a characteristic of an individual or a particular group, but of the entire church. Disability does not affect only certain individuals, but involves all of us together as the people of God in a broken world. It is our world that is shattered, and each of us comprises one small, fragile, and precious piece. We all hold the treasure of God's life in earthen vessels (cf. 2 Cor 4:7). Yet we hold it; and, what is more, we hold it together. In our attitudes and actions toward one another, at all times, the guiding principle must be the conviction that we are incomplete, we are less than whole, without the gifts and talents of all people. We are not a full community without one another. Responding to and fully including people with disabilities is not an option for the churches of Christ. It is the church's defining characteristic.

In relation both to the aged and the handicapped, we are faced with dimensions of being human, which affect us. Parents all over the world waiting for the birth of their child will every so often worry that their child will be born with a handicap Ageing is a condition which also encompasses fear at the end of life, fear of being deprived of health, intellectual capacity and hope for the future. Religion certainly functions as something to hold on to in situations of fear, worry and despair. There are those who would say that religion is nothing more than an attempt to seek comfort in anguish and anxiety.

The main thing for religion is not to yield to being reduced to becoming a "pacifier" for escapism but to enable people to voice distress and, at the same time, transform perceptions that being human includes being vulnerable and dependent, and that it is not a departure from the subject.

References

Biko, S. 2002. *I write what I like: Selected writings*. Chicago: University of Chicago Press.

Maalouf, A. 2000. *In the name of identity: Violence and the need to belong* (trans: Bray, B.). New York: Arcade Publishing.

World Council of Churches Central Committee. 2003. Geneva, Switzerland, 26 Aug–2 Sep 2003. http://www.oikoumene.org/en/resources/documents/wcc-commissions/faith-and-order-commission/ix-other-study-processes/a-church-of-all-and-for-all-an-interim-statement.html. Accessed 15 Aug 2011.

World Health Organization. 1980. *International classification of impairments, disabilities and handicaps*. Geneva.

Chapter 14
Health Care and Human Vulnerability: A Confucian Perspective

Jonathan Keung Lap Chan

14.1 Human Vulnerability

It is widely accepted that we bear special responsibilities toward those who are vulnerable, especially those who are vulnerable to our actions and choices. However, exactly who falls under the category of the vulnerable? What special responsibilities do we have toward them, and on what grounds?

These are the main questions this paper attempts to address. The question of who falls under the category of the vulnerable is a conceptual one. A satisfactory account of the category in question requires not only that it should be able to fix the extension of the 'vulnerable' but also that it be able to provide some reasons for ascribing vulnerability. Otherwise, 'vulnerability' functions as no more than a label. The question of what special responsibilities we have toward the vulnerable is different from the above question in nature.

This question is a moral one, whose answer depends on the truth or rational acceptance of some moral propositions whose justification in turn requires defending some moral perspective by means of which the truth or rational acceptance of the moral propositions can be explained. Some people attempt to explain the truth or rational acceptance of these moral propositions in terms of justice. In what follows, however, I shall argue that justice cannot fully account for the widely held moral intuition that we have special moral responsibilities to vulnerable people. As result, we need to go beyond justice and explore some other ground for it. In the paper, it will be argued that the Confucian moral perspective has ethical resources beyond justice to explain the moral intuition in question.

It should be noted, however, that my discussion will be confined to the context of health care and will reference to the Universal Declaration on Bioethics and Human Rights adopted by the General Conference of UNESCO in 2005.

J. K. L. Chan (✉)
Centre for Applied Ethics, Hong Kong Baptist University, Hong Kong, People's Republic of China
e-mail: klchan@hkbu.edu.hk

J. Tham et al. (eds.), *Religious Perspectives on Human Vulnerability in Bioethics*, Advancing Global Bioethics 2, DOI 10.1007/978-94-017-8736-9_14,

Some people are more vulnerable than others. Children, the elderly, women, the disabled and the severely ill are commonly deemed as the social groups which fall under the category of the vulnerable. But what enables us to ascribe vulnerability to these social groups? And what exactly is "vulnerability" meant?

One definition is that "vulnerability" means being susceptible of injury. At first blush, the definition successfully captures the ordinary meaning of the notion. It makes perfectly good sense to speak of children, for instance, being vulnerable because they are easily injured. Construed this way, vulnerability is then essentially a matter of being under threat of harm (Gooden 1985). However, this construal of the notion is deficient since not all kinds of vulnerability are constituted by threats. There are some kinds of vulnerability which are constituted by prejudices or unjust discriminations.

These latter cases, then, cannot be explained by the aforementioned definition. Accordingly, adopting the definition would result in a narrowly construed notion of vulnerability. A drawback of adopting the narrowly construed notion is that our moral focus would be obscured with regard to the ethics of vulnerability. Our moral concerns about vulnerable people would be focused merely on threats or physical harms to which these people are susceptible. The ethics of vulnerability would then be the ethics of preventing physical harms from befalling the vulnerable. However, for some people such as feminists, to understand the ethics of vulnerability this way is inadequate. A sound ethic of vulnerability, according to these thinkers, should take into account the kind of vulnerability constituted by social inequalities.

The phenomenology of human vulnerability and a broader moral concern about human vulnerability than preventing harms from befalling the vulnerable prompt us to seek for a less presumptuous account of 'vulnerability'. In what follows, I shall offer an alternative account of the notion which is, I think, less presumptuous.

My analysis of the notion draws on the idea of human capability which was first used by Amaryta Sen in assessing life quality, and developed by Martha Nussbaum in her political theory. Let us first give a brief explanation of what human capability means to Sen. On some occasions Sen explicitly defines "capability" as "the ability to achieve." (Sen 2000) However, the definition does not offer much help for understanding what Sen means by "capability" because he is not using the term "ability" in its ordinary sense. On some other occasions, Sen gives the following definition: "A person's 'capability' refers to the alternative combinations of functioning that are feasible for her to achieve."(Sen 1999, 2000, 2004) By "functioning" or "human functioning", he refers not only to voluntary action such as walking but also involuntary action such as breathing.

Sen extends the meaning of "functioning" to include a person's state of existence such as being healthy as well. Thus, a person's capability may involve not only voluntary actions such as walking and involuntary actions such as breathing but also states of existence such as being healthy. Two things should be borne in mind concerning the above definition. First, although a person's capability involves a set of functioning combinations, it does not follow that the person has to achieve or perform the functioning of the set. It is obvious that a person who is capable of killing does not have to kill.

Second, in most of the cases, a person's capability of doing something involves a set of her natural abilities which enable her to do that thing. However, the person's capability of doing that thing is not identical with the natural abilities set in question. For Sen, a person may have a set of natural abilities which enable her to walk, but lacks the capability of walking if that person is tied up by somebody. Or a person may have a set of natural abilities which enable her to vote in an election but lacks the capability of doing so if she is under age, or the law does not allow her to vote. Thus, women in a patriarchal society are less capable than women in a more equal society. That is why Sen refers not only to human functioning but also to the feasibility of a person's achieving those functions when defining "capability." The feasibility of achieving a certain set of functions involves not only a person's natural abilities but also the institutional arrangements of society.

Now let us turn back to the notion of vulnerability. By invoking the idea of human capability, it seems to be a straightforward task to give a satisfactory explication of the notion. The intuitive idea is that a person's vulnerability is constituted by her incapability. Suppose {C1, C2. C3...Cn} is a set of capabilities which are deemed as essential for a human adult to function normally. Then, we may define "vulnerability" as lacking some of the capabilities in the set. More capabilities in that set a person lacks, the more vulnerable she is.[1]

14.1.1 The Vulnerability of Children

Children lack the capabilities of fending for themselves, protecting themselves from injury, making right decisions and solving problems for themselves. In a nutshell, they lack the capability of leading an autonomous life. Obviously, these are the capabilities which a human adult should possess if she is to function normally, i.e., capabilities which should be included. It is exactly lacking these capabilities that makes children vulnerable. Children before birth are, then, even more vulnerable because they are less capable of fending for themselves and rely more on their mothers for their survival.

14.1.2 The Vulnerability of Women

Martha Nussbaum gives the following vivid description of the situation of women:

> Women in much of the world lack support for fundamental functions of a human life. They are less well nourished than men, less healthy, more vulnerable to physical violence and sexual abuse. They are much less likely than men to be literate, and still less likely

[1] Defining "vulnerability" in terms of a list of capabilities essential to the normal functioning of a person is not without problem. For people may have disputes about what functioning is to be regarded as normal and which capabilities as essential to the functioning in question. The disputes, however, cannot be easily settled in this short essay. In my view, what counts as "functioning normally" for a human adult and what capabilities are counted as essential to that normal functioning depend very much on the theory of the good one holds. In that case, the notion of vulnerability is a theory-laden concept. And any adequate treatment of moral issues in relation to human vulnerability must be conducted from a certain ethical perspective.

> to have professional or technical education. Should they attempt to enter the workplace, they face greater obstacles, including intimidation from family or spouse, sex discrimination in hiring, and sexual harassment in the workplace…All these factors take their toll on emotional well-being: women have fewer opportunities than men to live free from fear and to enjoy rewarding types of love—especially when, as often happens, they are married without choice in childhood and have no recourse from bad marriages. In all these ways, unequal social and political circumstances give women unequal human capabilities (Nussbaum 2000).

The above description of women's situation gives us some idea about the nature of the vulnerability of women. Women's vulnerability is largely due to their inferior status. It is their inferior status that makes them lack the capabilities which an average male adult in a democratic society has. However, besides the social and political arrangement, there is some condition under which women are deemed especially vulnerable, i.e., getting pregnant. We can make pregnant women less vulnerable by improving their social and political conditions. But it is hardly possible to eliminate completely vulnerability due to pregnancy.

14.1.3 The Vulnerability of Mentally and Physically Handicapped People

The vulnerability of the mentally disable can be explained in a way similar to the explanation of the vulnerability of children. The mentally disabled also lack the capability of fending for themselves, protecting themselves from injury, making right decisions and solving problems for themselves. That is to say, they lack the capability of leading an autonomous life.

Physically handicapped people are, however, vulnerable in a different way. The physically handicapped include blind people, deaf people, and those who have lost their limbs. The vulnerability of this group is mainly due to lacking the capability of performing certain bodily functions such as seeing, hearing and walking. Failing to perform those bodily functions makes them more difficult to control their living. This incapability, however, does not prevent them from leading an autonomous life. What they need is some assistance rather than the power of deciding their life decisions being taken over.

The disabled, whether mentally disabled or physically disabled, are vulnerable not only because they lose some abilities, mental or physical, but also because they are victims of discrimination. Like women, their vulnerability is partly due to the social practices or institutional arrangements in society.

14.1.4 The Vulnerability of the Elderly

Elderly people are often thought as weak, dependent, prone to confusion, and less capable than younger adults. This picture of the elderly, although it may not represent the whole truth of living in old age, captures the situation which many older people encounter. The fact of the matter is: declining health does prevent elderly people from retaining the degree of independence younger adults enjoy and, at some point in their old age, most need care. Due to declining health, they lack the capability of fending for themselves. Declining health is a major factor which contributes to the vulnerability of most elderly people.

Having severe illness is the next major factor which makes elderly people vulnerable. Elderly people who have severe illness, say, dementia, are even more vulnerable since they are less capable of fending for themselves than older adults who do not have severe illnesses.

Like the disabled, the elderly are vulnerable not only because their health is declining or because they have a severe illness but also because they are victims of discrimination. Elderly people who are capable of holding jobs often cannot get work not because jobs are scarce, but because there are prejudices against older workers. Therefore, the vulnerability of the elderly is also like that of the disabled. Their vulnerability is partly due to social practices or institutional arrangements in society.

14.2 The General Moral Position of the Confucian Perspective on Our Special Responsibilities to the Vulnerable

In the above section, I have examined a commonly used definition of the notion of vulnerability and argued that the definition is inadequate to capture the essential meaning of the notion. I then, by making use of Sen's idea of human capability, proposed an alternative definition of the notion and applied it to explain different kinds of human vulnerability.

Let us now turn to the question of whether we have special moral responsibilities to the vulnerable. In what follows, I shall give a brief description of the moral position of the Confucian perspective with regard to the above question. In the Confucian classic *Liji* or *The Classic of Rites*, there is a passage from the chapter entitled "Li Yun" which most of the Confucian scholars take to be representing the important social and political ideal of Confucianism. In that passage, Confucius attempts to explain the Confucian ideal of the Grand Union to his student Yan Hui by referring to the practice of the Grand course:

> When the Grand course was pursued…men did not love their parents only, nor treat as children only their own sons. A competent provision was secured for the aged till their death, employment for the able-bodied, and the means of growing up to the young. They showed kindness and compassion to widows, orphans, childless men, and those who were disabled by disease, so that they were all sufficiently maintained. Males had their proper work, and females had their homes… In this way (selfish) scheming was repressed and found no development. Robbers, filchers, and rebellious traitors did not show themselves, and hence the outer doors remained open, and were not shut. This was (the period of) what we call the Grand Union. ("Li Yun"in *Liji* or *The Classic of Rites*) (Confucius 1885)

According to the above passage, it is Confucius's view that the 'Grand Union' represents the morally ideal social world in which the practice of the Grand Course, i.e., the Confucian Way or Dao, prevails. In that world, people not only love their own parents and children but also care about the welfare of the people outside their family or clan. The elderly will be provided with sufficient food and other necessary things, and children with the means of growing up. Women will have their home, and widows, orphans, childless men, and the disabled will all be sufficiently maintained. In a nutshell, in the world of the Grand Union, not only the capable but also the vulnerable can lead a flourishing life.

From the above description of the world of the Grand Union, we can see that the Confucian Way requires that society be organized such that the welfare of the vulnerable is taken care of, so they can have a flourishing life as well. In other words, it is the moral position of the Confucian that we ought to take care of the welfare of the vulnerable and help them to achieve a flourishing life. But why does the Confucian think that we have such moral responsibility? For the Confucian, such moral responsibility can be better explained in terms of the Confucian notion of *ren* which is sometimes translated as humaneness or benevolence. In what follows, I shall explain what "*ren*" is meant for the Confucian and why the notion can help to explain the moral view that we have the moral responsibility to take care of the welfare of the vulnerable and help them to achieve a flourishing life.

On one occasion, Confucius gives a clear explication of the notion of *ren:*

> Tzu-Kung said, "If there were a man who gave extensively to the common people and brought help to the multitude, what would you think of him? Could he be called benevolent?"
>
> The Master said, "It is no longer a matter of benevolence with such a man. If you must describe him, 'sage' is, perhaps, the right word. Even Yao and Shun would have found it difficult to accomplish as much. Now, on the other hand, a benevolent man helps others to their stand in that he himself wishes to take his stand, and gets others there in that he himself wishes to get there. The ability to take as analogy what is near at hand can be called the method of benevolence." (*The Analects*, VI.30)

According to the above passage, Confucius takes a person of *ren* to be a person who possesses the moral capability of applying the "method of benevolence." The method of benevolence is a method which requires using analogical thinking of a certain kind, namely, making a moral analogy. To make a moral analogy, one expands one's circle of moral concern, extending one's own concerns onto others. In applying the "method of benevolence", one's concerns are not only one's own welfare but also others'. The possibility of such a kind of "analogical" thinking lies in the fact that, according to the Confucian, we have the moral capability of putting ourselves in someone else's situation.

It is exactly because of possessing this moral capability that we are able to see someone else's needs, sufferings and flourishing in the way we see ours.[2] It should, however, be noted that for Confucius, *ren* is not merely a descriptive concept. It is a normative concept as well. According to Confucius, *ren* refers to a virtue which

[2] This kind of "analogical" thinking can be found in Mencius as well. "Treat the aged of your own family in a manner befitting their venerable age and extend this treatment to the aged of other families; treat your own young in a manner befitting their tender age and extend this to the young of other families…In other words, all you have to do is take this very heart here and apply it to what is over there." (Mencius, IA.7)

a person ought to acquire or cultivate.[3] And once a person acquires the virtue, she "sees" both emotionally and cognitively that she ought to extend her own concerns onto others, including those who are vulnerable.

14.3 The Moral Position of the Confucian Perspective on Our Responsibilities to the Vulnerable in the Context of Health Care

In the above section, we have shown that it is the moral position of the Confucian perspective that we ought to take care of the welfare of the vulnerable and help them to achieve a flourishing life. We also, by invoking the Confucian idea of *ren,* have explained why the Confucian thinks that we have such a moral obligation. However, what exactly does such moral obligation require us to do?

For the Confucian, the answer to the above question must be context dependent. What exactly the moral obligation requires us to do varies from context to context. This is because for the Confucian, the moral obligation in question cannot be specified in terms of a set of concrete rules. They are to be determined in terms of the judgements of a person of *ren* according to contexts, assuming that her judgements are well-informed.[4] Thus, for the Confucian, our moral obligations are always moral obligations in a certain context.

In what follows, I shall discuss the moral position of the Confucian perspective on our moral responsibilities to the vulnerable groups in the context of health care. My discussion, however, will reference to the Universal Declaration on Bioethics and Human Rights adopted by the General Conference of UNESCO in 2005 (UNESCO 2005).

14.3.1 Protecting the Vulnerable

In October 2005, the General Conference of UNESCO adopted by acclamation the Universal Declaration on Bioethics and Human Rights. In what follows, my discussion of the Declaration will focus only on articles which are concerned with the vulnerable groups. Article 8 says:

[3] In The Analects, there are passages which state the normativity of ren. "The Master said, 'Of neighbourhoods benevolence is the most beautiful. How can man be considered wise who, when he has the choice, does not settle in benevolence?" (The Analects, IV.1) "If the gentleman forsakes benevolence, wherein can he make a name for himself? The gentleman never deserts benevolence, not even for as long as it takes to eat a meal. If he hurries and stumbles one may be sure that it is in benevolence that he does so." (The Analects, IV.5)

[4] It is the central idea of virtue ethics that rules cannot fully capture the complexities of the moral reality and, thereby, our moral responsibilities, and that only when one becomes a virtuous person is she able to grasp those complexities and know her moral responsibilities to others.

> In applying and advancing scientific knowledge, medical practice and associated technologies, human vulnerability should be taken into account. Individuals and groups of special vulnerability should be protected and the personal integrity of such individuals respected.

The Confucian would echo the above article that in applying and advancing scientific knowledge, medical practice and associated technologies, we ought to protect individuals and groups of special vulnerability, and respect their personal integrity, although on different grounds. For the Confucian, we have the obligation not because those individuals and groups of special vulnerability have a certain right.[5] We have the obligation in question because that is required by *ren*. The virtue of *ren* requires us to extend our concerns to others' elderly and children and show kindness and compassion to widows, orphans, childless men, and those who were disabled by disease.

And only when we fully develop the virtue of *ren* in ourselves, then we know exactly what we should do in order to protect the vulnerable individuals and groups in the context of applying and advancing scientific knowledge, medical practice and associated technologies. Special attention, however, should be paid to children, the mentally disabled and the elderly with severe illnesses such as dementia. This is because individuals in these groups lack the capability of protecting themselves from injury and making right decisions for themselves.

Accordingly, the Confucian would also echo Article 7 entitled "Persons without the capacity to consent" that special protection should be given to those who do not have the capacity to consent, in our case, children, the mentally disable and the elderly with severe illness such as dementia. As have been said above, exactly what to do with regard to the protection of the vulnerable groups depends on the judgements of persons who fully develop their virtue of *ren* in the relevant context. But the Confucian probably would endorse the clause (b) of the article: "research should only be carried out for his or her direct health benefit, subject to the authorization and the protective conditions prescribed by law, and if there is no research alternative of comparable effectiveness with research participants able to consent.

Research which does not have a direct potential health benefit should only be undertaken by way of exception, with the utmost restraint, exposing the person only to a minimal risk, minimal burden, and only if the research is expected to contribute to the health benefit of other persons in the same category…"

[5] For the Confucian, individual rights can never be the ground of morality. I have argued for this conclusion on many occasions. In brief, the notion of individual rights requires a social structure in which a sharp distinction between individuals exists and individuals are treated as equal. However, it is exactly such a sharp distinction between individuals and equality among individuals that is absent in the social structure which aims to develop people's ren. In a society with that social structure, how a person should be treated depends, though not entirely, on her relations to others. Equality, then, is relatively unimportant in the moral and social life of that society. Only against a social background which accords the state a distributive function could equality have an important role to play. Lacking such a social background, it would be meaningless to talk about equality unless what is being talked about is formal equality. However, in traditional Chinese society, the state was not accorded such a distributive function. And only when we understand this, are we able to understand why Mencius said, "That things are unequal is part of nature…If you reduce them to the same level, it will only bring confusion to the empire." Mencius (1984). 3A:4.

14.4 Against Social Discrimination

Article 11, entitled "Non-discrimination and non-stigmatization", of the Universal Declaration on Bioethics and Human Rights, states:

> No individual or group should be discriminated against or stigmatized on any grounds, in violation of human dignity, human rights and fundamental freedoms.

Although the article does not explicitly mention any vulnerable groups, it has a direct bearing on them. It is because the vulnerable are the ones who are usually the target of discrimination and stigmatization. The Confucian would endorse the article's saying that no individual or group should be discriminated against or stigmatized. To see this, we only need to apply the method of benevolence. In applying the method of benevolence, we are required to put ourselves in the situation of those who are discriminated against or stigmatized.

However, no one would want to be discriminated against and stigmatized. Thus, in applying the method of benevolence, we would agree that no individual or group should be discriminated against or stigmatized. Accordingly, the Confucian would also endorse the non-discriminatory principle implicit in the following part of Article 14

> 2. Taking into account that the enjoyment of the highest attainable standard of health is one of the fundamental rights of every human being without distinction of race, religion, political belief, economic or social condition, progress in science and technology should advance:
> (a) access to quality health care and essential medicines, especially for the health of women and children, because health is essential to life itself and must be considered to be a social and human good

In many areas of the Earth, women and children are usually the target of discrimination, especially in the context of health care. From the Confucian moral perspective, the discriminatory treatment of women and children is morally unacceptable. As we have seen, the application of the method of benevolence disallows us from adopting discriminatory practices against any individual or group in society. Thus, discriminatory practices against women and children in health care are deemed to be violating the principle of *ren* from the Confucian point of view.

14.5 Going Beyond Liberal Justice

The Universal Declaration on Bioethics and Human Rights states the following as one of the aims of the Declaration:

> to promote equitable access to medical, scientific and technological developments as well as the greatest possible flow and the rapid sharing of knowledge concerning those developments and the sharing of benefits, with particular attention to the needs of developing countries.

And Article 10 entitled "Equality, justice and equity" states

> The fundamental equality of all human beings in dignity and rights is to be respected so that they are treated justly and equitably.

Once again the above aim, and the article of the Declaration, do not mention any vulnerable groups explicitly. However, the two have direct bearing on the vulnerable as well. This is because the vulnerable are the weakest among those who compete for medical or technological resources in health care. And the above aim and the article of the Declaration do provide the vulnerable some moral protection of their interests in accessing the resources in question.

The Confucian would endorse the moral vision of the Declaration that all human beings, including those who are vulnerable, should be treated justly and equitably, and that we should promote equitable access to medical, scientific and technological developments, depending on how "being treated justly and equitably" and "equitable access" are to be understood. In contemporary social and political philosophy, distributive justice has dominated philosophical discussions on questions of justice ever since John Rawls published his *A Theory of Justice* in 1971. Since then justice and equity are mainly understood to be concerned with distributive justice. Article 10 has to be understood against this intellectual background as well. According to the article, treating a person justly and equitably requires that the fundamental equality of all human beings in dignity and rights be respected.

But what counts as "respecting all human beings equally in terms of their dignity and rights" depends on what theory of justice we hold. Thus, for the liberal such as Rawls, to treat individuals "justly" and "equitably" is to treat them in the way that the fulfillment of their needs is achieved by the institutions of the basic structure of society in ways the principles of distributive justice specify as fair:

> [G]iven the political conception of citizens, primary goods specify what their needs are… the specification of these needs is a construct worked out from within a political conception…what is really important in questions of justice is the fulfillment of citizens' needs by the institutions of the basic structure in ways the principles of justice, acknowledged by an overlapping consensus, specify as fair (Rawls 1993).

This Rawlsian conception of justice assumes that the major role of the principles of justice is to specify the conditions of the institutions of the basic structure of society so that distributions of the primary goods are fair; and therefore that the role of justice is mainly distributive. What justice aims to achieve is, then, to spell out fair social conditions under which the primary goods are distributed. These primary goods, according to Rawls, are rights and liberties, opportunities and powers, income and wealth, and the social basis of self-respect (Rawls 1993).

The above liberal conception of justice is, however, not without controversy. For some critics, it is general justice, which is concerned with the exercise of complete virtue, instead of distributive justice that should play the central role in exploring social justice (Aristotle 1985, pp. 116–127).[6] On these critics' view, the most impor-

[6] In Nicomachean Ethics, Aristotle distinguishes three subjects of justice, namely, general justice, distributive justice and rectificatory justice.

tant question of justice is how to organize the institutions of the basic structure of society in ways citizens can develop and exercise their complete virtue (Fan 2010, p. 47, 48).[7] Thus, according to these critics, treating a person justly and equitably requires treating her in ways the exercise of our complete virtue requires.

There is another problem concerned with the idea of distributive justice, even if we push aside the primacy issue. As some critics have pointed out, rival conceptions of justice exist, and there is no single standard of rationality to which our conception of justice can appeal (MacIntyre 1988, p. 1, 2).[8] Since there is no single valid standard of justice which can be used to judge whether a person is treated justly equitably or not, "treating a person justly and equitably" and "equitable access" become formal notions and are subject to interpretation. For the Confucian, "justice" and "equity" must be understood in terms of the ideas of *ren* and *yi*, which are the two supreme Confucian virtues, and distributions of other social goods have to be determined according to the social principles conducive to the development and exercise of the two supreme Confucian virtues.[9]

The third problem concerned with the liberal conception of justice is how to apply the idea of distributive justice to the vulnerable groups such as children and the mentally disabled, if distributive justice is understood in Rawlsian terms. The problem is: the mentally disabled are not, and children are not yet, normal and fully cooperating members of society, and therefore cannot satisfy the political conception of the person presupposed by the political conception of justice. According to Rawls, the political conception of the person requires that persons possess, to the requisite degree, the two powers of moral personality, namely, the capacity for a sense of justice and the capacity for a conception of the good

> [P]ersons were regarded as free and equal persons in virtue of their possessing to the requisite degree the two powers of moral personality, namely, the capacity for a sense of justice and the capacity for a conception of the good. These two powers we associated with the two main elements of the idea of cooperation, the idea of the fair terms of cooperation, and the idea of each participant's rational advantage, or good.[10]

It is obvious that children and the mentally disabled do not possess the two powers in question. Thus, these two groups of vulnerable people are outside the sphere of justice understood in Rawlsian terms. Now the remaining question is: how to explain the fact that most of us think that society has moral responsibilities to ensure

[7] "Confucians would agree with Rawlsians that there ought to be fundamental principles to direct the institutions…Confucians could not affirm that such principles primarily concern the distribution of primary social goods…If intrinsic goods such as ren and yi are not established in the first place, the concern with instrumental goods such as money or profit would not really do good."

[8] "Some conceptions of justice make the concept of desert central, while others deny it any relevance at all. Some conceptions appeal to inalienable human rights, others to some notion of social contract, and others again to a standard of utility…To know what justice is, so it may seem, we must first learn what rationality in practice requires of us. Yet someone who tries to learn this at once encounters the fact that disputes about the nature of rationality in general and about practical rationality in particular are apparently as manifold and as intractable as disputes about justice."

[9] Let us call these social principles the principles of Ren -Yi.

[10] Ibid. note 13, p. 34.

that the basic needs, including acquisition of adequate health care, of the two vulnerable groups are to be satisfied? To answer this question requires that our moral thinking goes beyond distributive justice or the liberal conception of justice. For the Confucian, the above moral intuition can be better explained in terms of idea of *ren*. As we have seen, in applying the method of benevolence, a person of *ren* would take it to be our moral obligation to take care of the welfare of the two vulnerable groups and help them to achieve a flourishing life.

Their basic needs will be taken care of not because they are normal and fully cooperating members of society but because of our sense of humaneness or *ren*.

References

Aristotle. 1985. *Nicomachean Ethics*, 116–127. Trans: T. Irwin. Indianapolis: Hacket.

Confucius. 1885. *The sacred books of China. The texts of Confucianism,* 364. Trans: J. Legge. Oxford: Clarendon Press.

Confucius. 1979. *The analects*. Trans: D. C. Lau. Hong Kong: Chinese University Press.

Fan, R. 2010. *Reconstructionist confucianism: Rethinking morality after the west*, 47–48. Dordrecht: Springer.

Gooden, R. E. 1985. *Protecting the vulnerable: A reanalysis of our social responsibilities,* 110. Chicago: University of Chicago Press.

Hall, D., and R. Ames. 1987. *Thinking through Confucius*. New York: State University of New York Press.

MacIntyre, A. 1988. *Whose justice? Which rationality?*, 1–2. London: Duckworth.

Mencius. 2003. *Mencius: A bilingual edition*. Trans: D. C. Lau. Hong Kong: Chinese University Press.

Nussbaum, M. 2000. *Women and human development: The capabilities approach,* 1. Cambridge: Cambridge University Press.

Nussbaum, M. 2011. *Creating capabilities*. Cambridge: Harvard University Press.

Rawls, J. 1971. *A theory of justice*. Oxford: Oxford University Press.

Rawls, J. 1993. *Political liberalism*, 181, 188. New York: Columbia University Press.

Sen, A. 1999. *Development as freedom,* 175. New York: Knopf.

Sen, A. 2000. The standard of living. In *The standard of living: The Tanner lectures on human values,* eds. A. Sen, J. Muellbauer, K. Kanbur, K. Hart, and B. Williams, 36, 140. Cambridge: Cambridge University Press.

Sen, A. 2002. *Inequality reexamined*. Cambridge: Harvard University Press.

Sen, A. 2004. Elements of a theory of human rights. *Philosophy and Public Affairs* 32 (4): 332–338.

UNESCO. 2005. Universal declaration of bioethics and human rights. http://portal.unesco.org/en/ev.php-URL_ID=31058&URL_DO=DO_TOPIC&URL_SECTION=201.html. Accessed 31 Oct 2012.

Chapter 15
Hindu Perspective on Vulnerable Groups: Children, Women, the Elderly and the Disabled

Vasantha Muthuswamy

15.1 Introduction

The Indian tradition dates back to 8,000 BC and has a continuous recorded history since the time of the *Vedas* for over 5,500 years. Several elements of India's diverse culture—such as Indian religions, yoga and Indian cuisine—have had a profound impact across the world.

India is one of the most religiously diverse nations in the world, with some of the most deeply religious societies and cultures. Religion still plays a central and definitive role in the life of many of its people. India is the birth place of Hinduism, Buddhism, Jainism and Sikhism, collectively known as Indian or *Dharmic* religions. Hinduism, the third largest religion in the world, is the main religious philosophy practiced by 80% of the people in the Indian subcontinent. Though by traditional definition Hinduism can be described as a religion, it is more accurately defined as a way of life (*Sanātana Dharma*).[1] *Dharma* means righteousness and good moral and ethical practices in accordance with the scriptures; includes all duties—individual, social and religious.

This "way of life" is mandated through a series of religious texts that describe certain rules by which one should abide.

The key to the individual and social ethics of Hinduism is this concept of *Dharma*, whose full implications cannot be conveyed by such English words as religion,

[1] The authoritative sources of Hinduism are divided into two classes: the *Srutis* and the *Smritis*. (Dwivedi 1988) *The Srutis* are eternal truths of religion heard by the sages and recorded for the benefit of others. These are the *Vedas* and the *Upanishads*. The *Smritis* are the secondary scriptures which expand and exemplify the principles of the *Srutis*. These consist of the Codes of law, the great epics *Ramayana and Mahabaratha*, the *Puranas* or *Chronicles* etc. The most important law givers are Manu, Yajnavalkya and Parasara. They give detailed instructions, according to the conditions of the time, to all classes of men regarding their duties in life. Hence their codes of laws are known as *Dharma Sastras*.

V. Muthuswamy (✉)
Indian Council of Medical Research, New Delhi, India
e-mail: vmuthuswamy@hotmail.com

J. Tham et al. (eds.), *Religious Perspectives on Human Vulnerability in Bioethics*, Advancing Global Bioethics 2, DOI 10.1007/978-94-017-8736-9_15,

duty, or righteousness. Derived from a root, which means to support, the word signifies the law of inner growth by which a person is supported in his present state of evolution and is shown the way to future development. A person's *Dharma* is not imposed by society or decreed by an arbitrary God, but is something with which he is born as a result of his actions in previous lives. *Dharma* determines a man's proper attitude toward the outer world and governs his mental and physical reactions in a given situation. It is his code of honour.

Thus *Dharma*, or righteousness, is the basis of both individual commitment and social responsibilities. Hindu philosophers emphasized personal ethics over social ethics. They argued that since society consisted of individuals, if individuals were virtuous, then social welfare would follow as a matter of course. Second, the general moral tone was very high in the ancient Hindu society, where everybody was expected to do his appropriate duties, which included, among other things, rendering help to one's less fortunate or vulnerable fellow beings. In the event of a conflict between individual and social *dharma*, the social *dharma* takes precedence. "He who understands his duty to society truly lives. All others shall be counted among the dead," declares Tirukural, a Hindu scripture in Tamil language.

Ethics, which concerns itself with the study of conduct, is derived, in Hinduism, from certain spiritual concepts and forms the steel-frame foundation of the spiritual life. Hindu ethics prescribes the disciplines for a spiritual life, which are to be observed consciously or unconsciously as long as humans live. Hindu ethics is mainly subjective or personal, its purpose being to eliminate such mental impurities as greed and egoism, for the ultimate attainment of the highest good. Objective ethics, which deals with social welfare, has also been considered by Hindu thinkers and is based upon the Hindu concept of *Dharma*, or duty. Objective ethics, according to the Hindu view, is a means to an end, its purpose being to help the members of society to rid themselves of self-centredness, cruelty, greed, and other vices, and thus to create an environment helpful to the pursuit of the highest good, which transcends society.

Hinduism speaks of certain universal ethical principles which apply to all human beings irrespective of their position in society or stage in life.

Hindu philosophy rests on three important doctrines—*karma*, *ahimsa*, and *samsara*. (Kuthiala 1994) *Karma* is widely defined as the set of deeds, good or bad, one "collects" throughout one's life. *Ahimsa* denotes non-violence or do-no-harm. *Samsara* refers to life now or in future. The doctrine of *ahimsa* mandates that all beings lead a life of nonviolence.

It is the practice of this cardinal virtue to the best of their ability that has made the Hindus what they are today. In a word, the pacific character of Hindu civilization is a result of this ideal of *Ahimsa*. The doctrine of *karma* refers broadly to all the acts one does throughout his or her life. Based on his or her *karma,* one can determine in what form one will be reborn in his or her next life (*samsara*). *Dharma* refers directly or indirectly to the set of obligations or duties enunciated by *Vedas* (and/or *Smritis*) which include religious, social, moral and legal obligations without discriminating sharply between them.

Dharmas in the sense of duties or obligations are classified by *Manusmriti* and other *Smriti* texts into two: *visheshadharma* and *sadharanadharma*. *Visheshadharmas* are specific obligations which are assigned to persons in accordance with their *Varna*, caste, gender, family status, stage in life etc. *Sadharanadharmas*, on the other hand, are common obligations which are assigned to persons irrespective of their *Varna*, caste, gender etc. These obligations are universal or universalisable. Hence it can be seen that Hinduism defines specific obligations to certain category of human beings who have to be treated differently when compared to the general population and this concept goes well with the issue of vulnerability since age (children and elderly), gender (women), *varna* (occupational groups), etc. have been considered under *Visheshashram* (special obligations).

This ethical distinction among groups has produced two main schools Hindu thought. Some contend that the structure of specific obligations should be followed as a fixed framework and common obligations may be followed without violating this framework. Others contend that the common obligations should be followed primarily by all and specific obligations may be interpreted and modified in such a way that common obligations are not violated. We find that the Vedic culture generally emphasizes the structure-specific obligations and prescribes the pursuit of common obligations within their framework.

Hindu society recognises and values interdependence. According to Vedic theology, society can meet everyone's legitimate needs if the various individuals perform their respective duties. These duties embody the ideal of extending God's shelter to others. For this purpose, the system of varnashrama-dharma allocated specific duties to each *varna* and *ashram*. The Varnashrama-dharma defines duties for the individual, classified according to four divisions of labour and four stages in life.

These specific duties change, for example as one passes through the different ashrams or stages. Varnashrama-dharma is the basis for accommodating diversity, and attributing different social and spiritual standards to various sections of society. Although varnashrama-dharma relates largely to social matters, it is not divorced from sanatana-dharma but is a means of recognising a common goal approached from different starting points. The Vedic (Hindu) view is that execution of one's duties automatically fulfills the rights of others, and that stressing *dharma*, fostering a climate of social and spiritual responsibility. The Hindus acknowledgement of interdependence thus differs from the individualistic, self-centred approach to life.

Hence it is evident that Hinduism is basically a "Duty based" rather than "Right based" religion. Dr. S. Radhakrishnan, one of the greatest philosophers and former President of India, defines 'The Concept of Dharma' in his book on Religion and Society. He says, "The principles which we have to observe in our daily life and social relations are constituted by what is called dharma. It is truth's embodiment in life, and power to refashion our nature." (Radhakrishnan 1947)

In the context of the present conference on human vulnerability and bioethics, the four vulnerable groups *viz*., women, children, elderly and the handicapped who are vulnerable due to their reduced autonomy as a result of gender, age, physical and mental challenges, also come under the special category of those requiring *vishesh-dharma* or special obligations from the others. According to Hinduism, these groups of individuals always need protection by the others in order to carry out their responsibilities in life. Hence in the context of research, they are not autonomous beings who can decide for themselves what is good or bad but need social protectors or benefactors to assist them in decision making. The human extended family is called to care for their less fortunate members. It is interesting to examine how Hinduism has looked into these four groups and provides them with sufficient care and support throughout their life through responsibilities entrusted to other members of the family and society.

15.1.1 Women

Of all the organized religions of the world, women have perhaps the most prominent presence, both visible as well as invisible, in Hinduism. As the Divine Mother, the Supreme Being affirms to Hindus that It either has no gender, or It has both. Hindu women have sustained *Dharma* in various ways down the ages and played a central role in society. When God is worshipped as 'Divine Couple' by Hindus, the name of the feminine typically precedes that of masculine. '*Sita-Ram*', '*Radhe-Shyam*', '*Uma-Mahesh*' or '*Shri Vishnu*' and so on. (Sharma 1988) In Hindu *Dharma*, God is often compared to a mother, and is worshipped in the form of the Divine Mother.

In social contexts as well, no person is considered as exalted and worthy of respect and service as one's mother. The tender love and care of a mother for her children is the subject of numerous Vedic verses.

There was no restriction in the ancient Hindu society that sons must always be named after their father. Numerous heroes of Hindu tradition are frequently addressed as sons of their mother. Pregnant women were granted many exemptions due to the high regard for motherhood in the traditional Hindu society. The sanctity of motherhood was so highly regarded that it was also extended to the animal kingdom. It was forbidden to hunt pregnant animals. Amongst the most important reason for considering the cow a sacred animal in Hinduism is the exceptional motherly love, patience and concern that she exhibits towards the newly born calf—a scene used frequently in the metaphors of Vedic hymns.

If we study the ancient history, we find that women held top religious and social positions in the Vedic period. There are references to women sages and saints in the *Vedas* and the *Upanishads* who were greatly revered for their religious and spiritual wisdom. During and following the epic period, the caste system (an ancient social philosophy) became rigid, which caused conflict within the society. Women often became the victims of this internal social conflict, as well as of the violence caused by foreign invaders. The protection of women thus became a pressing issue for the society and men had to shoulder this responsibility.

Further, as in all human cultures, at some point of time the Hindu culture also unfortunately showed a preference for the male child. There are pre-natal rites prescribed by Hindu texts to ensure that the fertilized embryo is male and not female. Nonetheless, Hindu texts say that the daughter deserves compassion from her parents, and is the highest object of her father's compassion. It was forbidden to inflict physical punishment on one's daughter. All forms of infanticide and abortions are considered heinous sins and a true Hindu can never kill his own daughter, who is the very embodiment of Devi, and the object of reverence as well as of supreme compassion.

No wonder then, Hindu and Sikh religious leaders have openly come out against this heinous practice which has no precedent and is sanctioned against in Hindu *Dharma*.

Hindu *Dharma* is perhaps unique in having a 'brother-sister' festival, called the *Rakshabandhan* for reminding the brothers that they need to protect their sisters at any cost. On this day sisters come to visit their brothers and tie them a sacred thread (called '*Rakhi*') on their wrist to symbolize that if ever some adversity befalls upon them, their brothers will rise to the occasion. Sisters pray for their brothers' welfare, and brothers gift something to their sisters. In the Hindu tradition, a woman can make someone else a 'brother' by tying the *Rakhi* to him, and all the requirements and duties of a brother towards his non-biological sister are then expected to be fulfilled by him. Hindu texts are unanimous in declaring that God does not differentiate between men and women. From a Dharmic perspective, adherence to *Dharma* alone decides who is great and who is lowly.

Hindu society has produced numerous women who were able rulers, warriors, poetesses, scholars, mathematicians, freedom fighters, musicians, artists and so on.

The *Vedas* consider the wife as auspicious, the most auspicious one. She is the light of the home, the harbinger of many blessings, and worthy of great honour. The *Mahabharata* says the wife is her husband's best friend. However, social mores have prevented many Hindu women from realizing their full potential. In the Hindu society, women have been often subjugated in their various roles.

Evil customs such as prevention of widow remarriage, infatuation with having male progeny, etc. have added to the misery of millions of Hindu women. Although women may be classified according to *varna*, the *Manu Smriti* talks of three stages for a woman: The critical need to protect the women during the ancient period is clearly reflected in the following verse of the *Manu Smriti* (MS):

> Father protects (her) in childhood, husband protects (her) in youth, and sons protect (her) in old age. A woman cannot be left unprotected. (MS 9.3)

1. As a child protected by her father: Traditionally, girls did not receive a formal academic education. A woman's role, considered essential in preserving social and cultural values, was learned in the home.
2. As a married lady, protected by her husband: Hinduism places great value on pre-marital chastity and this has significantly influenced practices. Girls were betrothed and married at a very young age. In married life, the wife's roles were centred on the home and she was not burdened with contributing towards the family income. Fulfilling one's responsibility as a loving and available parent was considered paramount.
3. As a widow, protected by the eldest son: If the husband died or took to *sannyasa* or detachment from the worldly life, then the widow would be looked after by the eldest living son. Elderly women were always treated with great respect.

According to tradition, women, more delicate than men, require and deserve protection. Hindu texts extol the virtues of womanhood and of the essential role women have in nurturing future generations and the need to protect them in different stages of life. However, woman has the same religious and spiritual freedom in Hinduism as man. Like a man, she is the soul in bondage and the goal of her life is the same as that of man, spiritual perfection or *moksha* through selfless work, meditation and yoga. However, the status of women in Hindu society has been affected by factors other than the ideals set forth in the *Vedas* and the *Upanishads*, such as cultural mores and the exploitation of the biological and psychological differences between men and women.

Therefore, on an individual and social level, complete and total equality of women is a goal that Hindu society (and other societies) is still striving for. As Swami Vivekananda says, we must realize that man and woman are two wings of the same bird; that in order to truly soar to great heights, a man and woman must work in unison in order to achieve greater harmony in life.

> Where women are honored, there the Gods are pleased. But where they are not honored, no sacred rite yields rewards. (MS 3.56)

15.1.2 Children

Hindus love their children dearly, considering them gifts of God and products of their previous karmas. According to Manu, man reveals himself through his children. (Srinivasa Murthy 2010) Since a Hindu firmly believes in rebirth he views his own life from a wider perspective that encompasses not just this life but many other lives that precede and succeed it, and his individual existence as part of the cosmic cycle. Since ancient times, the main purpose of marriage was to raise children. They were important not only in their own right, but also for continuation of the family lineage, and the performance of the last rites for parents.

In some circles, nurturing pious and emotionally stable progeny was considered a valuable socio-spiritual contribution.

A son is generally preferred because he upholds the family values and ensures its continuity. A girl child is considered as '*parayadhan*' or someone else's property as she has to be sent to her husband's family after marriage. Hence, in many families girl children are subjected to gender bias. Hindus are very possessive about their children and spend a great deal of time and energy in bringing them up for the welfare of the child, many rites/rituals are performed at different stages of life starting from preconception for a woman to conceive, then at 3 months pregnancy to invoke Gods for a male child, at the time of birth, 6 months after birth when solid food is introduced, a hair cutting ceremony, at the completion of 1st year etc.

Children were also frequent recipients of welfare measures. The traditional Indian view of child welfare is based on *daya, dana, dakshina, bhiksha, ahimsa, samya-bhava, swadharma and tyaga*, the essence of which were self-discipline, self-sacrifice and consideration for others. It was believed that the wellbeing of children depended on these values.

As per *Bhagavad Gita*, thc hcad of cvcry Hindu family is required to take care of children until they settle in life. The time period for the same is not an open factor. For a boy the considered age is around 25 years. For a girl, the responsibility of the family is primarily until she gets married. The children are then expected to take care of their parents and grandparents in a joint family system.

Hence parents are custodians of children till they settle in life and once settled, the young adults take care of their parents ensuring a bond within the family. Hindu thought divides life into 25 year stages, giving specific ethical advice for each: the first stage is for learning (*Brahmacharya*), the second is the time for the house holder (*Grahasthashrama*), the third stage is a time for meditation and study of scriptures (*Vanaprastha*) and the final stage is one of renunciation of the outer life (*Sanyasa*). The sequence ultimately ends in libration, the goal of life. The parents slowly detach themselves from the family in the *vanaprastha stage* (stage of detachment) and then move on to the *Sanyasa stage* (stage of total denouncement from the worldly desires) to complete the life cycle in this birth.

15.1.3 The Elderly or the Aged

The aged in India have been highly regarded since the earliest times of Indian history. Teachings of Hinduism strongly emphasize the concept of filial piety along with a moral responsibility to take care of one's parents in old age. (Tiwari 1998) Hindus are expected to venerate and worship their ancestors ritualistically.

A period of *shradhs* among Hindus is set aside each year to engage in such rituals to perpetuate the memory and remembrance of the deceased parents. If one worships their ancestors, they are blessed and prosper in their chosen profession or business. Family plays a significant role in the Indian culture. For generations, India has had a prevailing tradition of the joint family system under which extended members of a family—grandparents, parents, children, the children's spouses and their offspring, etc.—live together.

Usually, the eldest male member is the head in the joint Indian family system. He makes all important decisions and rules, and other family members abide by them. The patriarch often resolves family issues. Indian culture emphasizes the reverential treatment of the aged persons. The social system puts pressure on the children, especially the male children to take care of the aged parents, apart from the legal requirements.

The cultural practices also assign certain duties for the aged in the household chores. Taking care of the young children, looking after the societal responsibilities, settling inter-personal or inter-household or even inter-group conflicts, helping in the matrimonial match-making, are among the duties that society expects the elderly to take interest in and attend to. Thus, the aged are made to play useful roles in the household and in the society so as to make them feel reassured that they are an important part of society. They also transmit cultural values to younger generations through educational and entertaining storytelling.

In a joint family not everyone is hard working or successful in their endeavours. By the efforts of those who are energetic and able, the older parents as well as less capable kin are supported with a sense of moral obligation, religious duty and filial piety. By doing so one is able to perform both the Dharma and Karma expected of all Hindus transmigrating a cycle of births and rebirths in this world. The Law of Karma is one of the fundamental doctrines of not only Hinduism, but also of Buddhism and Jainism. As a man sows, so shall he reap. This is the law of Karma.

Desire produces Karma. You work and exert yourself to acquire the objects of your desire. Karma produces its fruits as pain and pleasure. You will have to take births after births to reap the fruits of your Karmas. This is the law of Karma. The doctrine of reincarnation or transmigration is a fundamental tenet of Hinduism.

It is believed that reluctance to take over such obligations or perform them according to the social norms could result in invisible, unexpected and sometime heart wrenching consequences such as severe illness, birth of a handicap child, injury to one's self, destruction of a valuable possession, or severe mental anguish. There are overwhelming cultural pressures to take care of elderly parents or next of kin. Taking care of one's parents in old age is considered to be one of the most important virtues that a person can possess. The elderly still hold property rights, act as patriarchs, are considered an infinite source of wisdom, and serving them improves one's Karma and Dharma.

We can expect modernisation to have a slow, but steady impact on the status of the elderly both in urban and rural areas in the next century. The extended family system continues to provide a functional division of labour thereby making the elderly useful to the family. Hierarchical relations assure the elderly a reasonable level of respect and self-dignity most situations and circumstances. Yet, respect for the older order may be declining and prescribed roles for each stage neither believed nor adhered to. Hindu social order, which has provided both stability and continuity to the Indian civilisation for thousands of years, is undergoing an intense scrutiny.

With its genius for synthesis and a very strong ability to accommodate one can only hope that, even with new directions, the majority of the elderly would be cared for by the family.

15.1.4 The Disabled

Hinduism is the oldest religion of mankind and contains an ocean of spiritual teachings about all aspects of life and consciousness. Mental health is an important component of Hinduism in addition to physical health. Hinduism focuses on five areas of mental health.

First, spirituality is an essential feature of the practice of Hinduism. Second, there is emphasis on unity of the body and mind. Third, yoga and meditation provide a practical way of addressing health in general and mental health in particular.

Fourth, *Bhagavad Gita*, one of the key texts of Hinduism, is an excellent illustration of the application of psychological interventions to life situations. Finally, the concept of *ashramas* (stages of life) provides a guide to master the physical and mental changes during the life-course of individuals.(www.archaeologyonline.net/artifacts/scientific-verif-vedas.html)

Bhagavad Gita is one of the most translated religious classics in the world. The beauty and sublimity of the work, the eternal relevance to the problems of human life and its universal approach that helps to consider the whole creation as one are the special features of this important text. It presents a detailed account of human emotions and cognitive deviations. The human mind and its weaknesses are elaborated. There is an emphasis on how to gain mastery over the vacillating mind.

The consequences of failure to attain such mastery are illustrated. Having understood these, there is a method to overcome these cognitive distortions as well as attaining full control over the mind so that a person can be his/her own master.

In *Ayurveda*, the science of life and health, the fundamental principle of health is the proper balance between five elements (*Bhutas*) and three humors (*Doshas*). (http://www.atributetohinduism.com) The balance occurs at different levels: physical, physiological, psychological and finally spiritual—the state of bliss in which the ultimate goal is tranquillity. All the major Ayurvedic texts like *Caraka Samhita* and *Susruta Simhata* have a separate section dealing with insanity (*unmada*). *Ayurveda* describes diseases resulting from the excessive use (*atiyoga*), deficient use (*ayoga*), or improper use (*mithyayoga*) of the mind. These diseases are known to result from the misuse of intelligence. Insanity 'is the unsettled condition of the mind, understanding, consciousness, perception, memory inclination, character, behaviour and conduct'. The symbiotic relationship between '*psyche*' and '*soma*' was recognized in *Ayurveda*, attributing the highest importance to psychic energy as the propulsive power of creation—the original force.

Man was treated as a whole with a psychosomatic approach. The therapeutic measures for insanities vary from words of sympathy and comfort to terrorizing means. Hinduism and other religions allied to it champion the concept that deeds performed in past births affect subsequent births. This concept, known as *karma*, has offered a powerful explanatory model: thus, a mentally sick person is supposed to bear the burden of misdeeds he must have performed in his past lives.

Suffering is inevitable. It is in this way that one can neutralize the effects of bad *karma*. This ideology built upon the will of God and *karma* generates paradoxically two views: first, one may be indifferent to a person suffering from illness (mental illness included) because his suffering is a pronouncement of divine justice, therefore inevitable. As a consequence, this leads to ridiculing mental patients. Families may also break their ties with those having mentally ill members. The family of the mentally ill is bound to become a 'social island' in the process.

The other attitude this ideology generates is one of compassion. If one casts aspersions at mentally ill patients, one accumulates bad deeds, for which he shall have to undergo divine malediction in the next birth. Therefore, one should sympathize with the ill and their families and pray that they may be cured. Similarly the physically handicapped are made to suffer in this birth because of their bad *karma* in the previous birth, and hence need compassion for the suffering their misdeeds merited. While no one can escape from the karmic effects, it is the *dharma* (duty) of others to take care of such individuals so that they themselves don't fall a victim to the karmic cycle.

15.2 Conclusion

This article has examined Hindu perspectives of four different vulnerable groups: women, children, the elderly and the disabled, with particular attention paid to how the concept of *visheshdharma*, or special obligations, makes it mandatory for the other able members of the family and society to care for them. The paper does not delve into the issue of biomedical or health research, but rather examines the *visheshdharma* enshrined in the Hindu scriptures for protection of the aforementioned groups discussed in this workshop on Human Vulnerability.

> Persons who render selfless service to other human beings are the greatest. Persons who carry on their profession, avocation or business with self interest, but without exploiting and causing any injury to those who deal with them are good. But those who give trouble to or exploit others in utter selfishness are demons in human form. (*Nitishataka* 72)

References

A Tribute to Hinduism. 2008. A beautiful website set up by Sushama Londhe, a Hindu woman, as a labor of love. http://www.atributetohinduism.com. New Delhi: Pragun publications.

Besant, A., and B. Das. 2000. *Sanatana dharama: An advanced textbook of Hindu religion and ethics*. Chennai: Theosophical Publishing House.

Dwivedi, L. D. 1988. *Introduction to Ayurveda*. Varanasi: Krishnadas Academy.

Kuthiala, S. K. 1994. Caring for the elderly: New dimension in India. *Health and Population—Perspectives & Issues* 17 (1–2): 86–96.

Radhakrishnan, S. 1947. *Religion and society*. London: George Allen and Unwin Ltd.

Sharma, D. S. 1988. *Essence of Hinduism*. Bombay: Bharatiya Vidya Bhavan.

Srinivasa Murthy, R. 2010. Hinduism and Mental health. In *Religion and psychiatry: Beyond boundaries*, eds. P. J. Verhagen, H. M. van Praag, J. J. Lopez-Ibor, J. L. Cox and D. Moussaoui. Wiley: New York.

Tiwari, K. 1998. *Classical Indian ethical thought*. Delhi: Motilal Barsidass.

Visnu, S. 2005–2013. Scientific verification of Vedic knowledge. Archaeology online. www.archaeologyonline.net/artifacts/scientific-verif-vedas.html. Accessed 15 june 2011.

Chapter 16
Islamic Perspectives on Vulnerable Groups

Dariusch Atighetchi

16.1 Assault and Invasion?

Very few people realize how contemporary medicine and the technology associated with it represent powerful instruments not only of technological but conceptual standardization derived from culturally specialized model(s). Different conceptions of the human body, numerous in past centuries, increasingly tend to be monopolized by the progressive standards of a particular culture and to the detriment of other conceptions that have disappeared or are disappearing. The West and other technologically advanced countries live in a state of "permanent scientific revolution" drawing with them, thanks to globalization, all other countries, obviously to a different extent and depending upon local situations (Atighetchi 2007).

The first impact of modernity is the technological products that are received.

The influence of modern Western-international normative models can be a form of legal-intellectual colonization. Western influence appears in many different ways and is present in almost all legal frameworks of Muslim states. Most of them have implemented modern Western public laws. Since the middle of the nineteenth century, models of Roman law (and to a lesser extent the Common Law) has predominated and become the point of reference for juridical acculturation in Muslim states. All these states have become legislators, including Saudi Arabia which nevertheless asserts (with respect to the *Shari'a*-based approach) that its legislative activity is only an application of the *siyāsa shar'iyya* (i.e. a mere administrative activity aimed at applying the *Shari'a*) rather than an actual autonomous activity producing legal norms (Castro 1990). The *Shari'a* has never been codified.

Codification started in the last two centuries under Western influence with the imitation of codified juridical models from the West. It is now dominant in Muslim states and the rest of the world. Imitation of civil (Castro 1985), criminal, commercial and administrative "Western" codes can be seen almost everywhere: organization of collective courts of law (instead of the traditional *qadi*, the monocratic

---

D. Atighetchi (✉)
Islamic Bioethics, Faculty of Theology, Lugano, Switzerland
e-mail: dario_atico@iol.it

J. Tham et al. (eds.), *Religious Perspectives on Human Vulnerability in Bioethics*, Advancing Global Bioethics 2, DOI 10.1007/978-94-017-8736-9_16,

judge); different levels of jurisdiction; the introduction of a Supreme Court; the introduction of a constitution (a classic Roman legal institution). It is not surprising that Saudi Arabia was the last Islamic country to introduce a constitution into its legal system after a fierce internal debate by Muslim jurists.

The greatest resistance to juridical innovations (*bida'*) has always come from family and inheritance laws since these areas are inspired directly by the Koran and the *Sunna* and are therefore more difficult to "modernise" or modify (Aluffi Beck-Peccoz 1990). At present, any new application of *Shari'a* based rules (e.g., in criminal law) takes place through the laws of the state. In practice, the very Islamization or re-Islamization of a state takes place mainly through juridical instruments of other traditions.

Transformation has been so considerable that Muslim law, due to external normative and cultural influences, changed from being extra-territorial, personal and confessional to state and territorial law (Castro 1981). The tripartite division of Islam into a juridical, religious and political system reflects a Western analytical-juridical based pattern. In turn, Roman categories that distinguish dogma, morals, rites, public and private law are ill-adapted to reflect the uniqueness of Islam as in the phrase "*Din wa Dawla,*" i.e., "Religion and State." As a partial counterbalance and in response to these external influences, some Constitutions list certain fundamental human rights while, at the same time, allow for some specific rights only if they are compatible with the *Shari'a*. This mechanism allows the reassertion of the latter's primacy over every other juridical approach. (e.g., art. 26 of the Saudi Arabian Constitution states, "The state protects human rights in accordance with the Islamic *Shari'ah*").

In addition, international declarations and documents on human rights, women's rights, and children's rights, etc., are paralleled with a series of documents on the same subjects aimed at reasserting Islamic identity and specificity. Of particular importance are the Declaration of Human Rights in Islam approved in Cairo by the Organisation of the Islamic Conference (OIC) on 5/8/1990 (International Human Rights 1993) and the Universal Islamic Declaration of Human Rights promulgated in Paris at a UNESCO meeting on 19/9/1981 by the Islamic Council of Europe.

Principal "Western" universal declarations appear "particular" because of their intrinsically secular nature or religious neutrality. On the contrary, for Islam, only the Law revealed by the Creator (the *Shari'a*) is truly universal, a universality that cannot be shared by laws founded exclusively or mainly on human reason.

16.2 Introduction to Vulnerabilities

Without further entering into the conflict (at least partially) of universalisms, we will now discuss the "East-West" relationship on the themes of vulnerability. The prevalently Western standardizing influences have disrupted or manipulated the numerous conceptions of infancy, old age, women and the handicapped which have been present in the different cultures formed by social, cultural and juridical processes over thousands of years.

Each of the four (very broad) topics on vulnerable groups will be discussed in a limited and particular perspective. The approach taken will look at the problems and is more constructive rather than apologetic. Although inevitably brief, we will address the following sections: (1) "*Children before and after birth*" will focus on the subject of infusion of the soul; (2) "*Women*" will focus on the subject of acknowledgement and application of free and autonomous consent by women prior to undergoing clinical tests, diagnosis and therapy; (3) "*The physically and mentally handicapped*" will look at some elements in the case of mental illness from an Islamic perspective; (4) "*The elderly*" will provide a view on the provisions of "homes" and palliative care for the elderly.

16.2.1 Children Before and After Birth

The infusion of the soul by God is one of the most important factors influencing the management of infantile health before and after birth. The legal status and vulnerability of children before and after birth has always been greatly influenced by these juridical interpretations.

The Koran and the "sayings" (*ahadith*) of the Prophet contain many references to the development of the embryo. The value of human life can be inferred from many passages in the Koran[1] (e.g. 17:31; see also 81:8; 4:29; 5:36), and this value is protected within the limits and exceptions laid down by the *Shari'a*; furthermore the Koran and *ahadith* prohibit infanticide (e.g. 6:137; 6:151; 6:140).

Of the passages in the Koran which God and the Prophet Muhammad speak of the embryo, the most exhaustive is found in Koran 23:12–14, which describes seven stages of the development of the embryo in the maternal womb: "12. Man We did create from a quintessence (of clay); 13. then We placed him as (a drop of) sperm [*nutfa*] in a place of rest, firmly fixed; 14. then We made the sperm into a clot of congealed blood [*alaqa*]; then of that clot We made a (*foetus*) lump [*mudgha*]; then We made out of that lump bones and clothed the bones with flesh; then We developed out of it another creature."

The last phrase of 23:12–14, "a new creation", introduces the most characteristic element of Muslim thought, i.e. the ensoulment of the foetus by God. Ensoulment distinguishes two phases in the development of the foetus, the first without a soul and the second with a soul: this distinction has taken on great relevance in the attempts to determine the juridical value of the different abortive acts.

As for the moment in which ensoulment takes place, such information is provided by various *ahadith* present—amongst others—in the authoritative collections of Bukhari, Muslim, and Nawawi. According to Bukhari, the Prophet Muhammad

---

[1] The Holy Koran (1983).

says: "The germ of every one of you is concentrated in his mother's womb in the form of a drop for 40 days; then he becomes a clot of blood for the same period; then he becomes a piece of flesh for the same period; then the angel is sent to him to ensoul him." In conclusion, the *ahadith* indicate different times of ensoulment, the most common being 120 days from fertilization, with other earlier and less common estimates, e.g. 40 days after fertilization.

The Muslim views on abortion are historically complicated and diversified, due to the criteria of infusion of the soul and the stages of development mentioned in the Koran and *Sunna*. These criteria form a "step-like" structure both at the level of juridical thought and at the penal level. It is difficult to summarise such a variety of positions on abortion in Islam. There has always been agreement in prohibiting (*haram*) abortion procured after infusion of the soul (120 days after fertilization or an earlier date), except in the case of therapeutic abortion. On the other hand, opinions on abortion before infusion are historically contradictory.

Contemporary positions oscillate between two poles which in turn can be divided into sub-groups. The overall picture obviously resumes the complex elaborations developed by the doctors of the Law in previous centuries. (1) A rigid pole which accepts only therapeutic abortion; in turn therapeutic abortion can be allowed only before infusion of the soul (the strictest version within 40 days, or within 120 days, etc.) or even after ensoulment of the foetus. (2) A moderate pole that accepts abortion not only to save the mother's life but also to protect her psychic and/or physical health and/or in the case of a handicapped foetus.

Regarding the laws of contemporary Muslim states,[2] laws based on the embryonic stages of Koranic derivation seem to offer—in principle—less protection for the embryo compared to those that ignore the embryonic stages but prohibit abortion from fertilization. This happens possibly because the presence of intermediate stages in the development of the embryo favours diversified responses to abortion at the various stages.

In general, there is no unanimous opinion on the use, experimentation and destruction of surplus fertilised eggs. Therapeutic research on embryos in the early phases is generally accepted with the parental consent. The criterion of infusion of the soul can be decisive to permit experimentation or use of pre-embryos before ensoulment.

Many Islamic legal and medical authorities uphold embryo stem cell research during the first days of the embryo (that is, long before ensoulment) with the aim of improving the public benefit (*maslaha*). In these cases (Atighetchi 2009), the legal principle of the public benefit (*maslaha*) overrides the protection of beings without a soul and living outside their mother's womb (i.e. "*in vitro*").

Generally, the advocates of this line of thought hold that the destruction of supernumerary embryos does not affect real human beings. As a consequence, research on soulless embryos can be legitimate. There is unanimous agreement on the requirement of consent from parents from whom the embryo originates and prohibition against producing embryos exclusively for the purpose of creating stem cells.

[2] As for the laws of Muslim states on abortion and their relations with the positions of the local juridical-religious authorities, see *Atighetchi D.* Islamic Bioethics: Problems and Perspectives. pp. 111–133.

There are also physicians and doctors of Islamic Law that believe the embryo must be protected from the first stages of development and that embryonic stem cell research should be considered illegal.

16.2.2 Women

Verbal mediation or the simple presence of a male relative (starting from the husband) before a male doctor are frequent requests amongst female Muslim immigrants in Europe. This mediation appears in many contemporary cultures to be an offence against the equality and dignity of women. In this regard, Muslim families who uphold this habit do not judge male mediation as an abuse of women's rights. For example, Saudi Arabian women appreciate the social security the family provides, given that the male has the legal and cultural responsibility to meet the woman's needs regarding food, clothes, home, health needs, etc.

However, this social relationship could limit a woman's rightful independence and autonomy. For instance, some male relatives sign the consent form even though it is their women's right to make independent decisions. The role of male tutors tends to diminish the autonomy of the woman under their protection. Furthermore, the protector's role tends to increase as the woman's health increases in gravity. The role of the family in protecting the ill woman may risk limiting her autonomy and freedom of treatment.

As we can easily imagine, Muslim women, especially in traditional areas, are reluctant to be examined by men even in the presence of another woman[3]. In this regard, the Muslim Law Academy of Mecca (a body of the Muslim World League) pronounced at the 8th session, on 19–28 January, 1985. The document[4] states that a Muslim woman is prohibited to undress in front of a man with whom it is unlawful for her to have sexual intercourse. This is allowed only for legitimate purposes recognized by the *Shari'a* which includes the case of a woman who requires medical treatment or care. Even in similar circumstances, however, the doctor should be a Muslim woman or, alternatively, a non-Muslim female doctor.

In the absence of both, the doctor must be a male Muslim or, in the absence of this, a non-Muslim. It is however prohibited for the male doctor and the female patient to be alone and the husband, a male relative (*mahram*) or any other woman must be present. Based on the usual Islamic legal "realism", in the case of grave necessity, these restrictions are widened (e.g. in births, orthopaedic fractures etc.) Hence, the doctor must treat the woman whose life is at risk, even when this infringes on the aforementioned principle out of the necessity to save human life.

Due to an emphasis on the separation of the sexes, doctors and nurses should refrain from touching a patient of the other sex except during treatment. Regarding intimate examinations, art. 11.1.9 of the Code of Ethics (2001–2002) of the Pakistan Medical and Dental Council states that: "For any intimate examination, the patient

[3] Atighetchi (2007, pp. 43–46).

[4] Atighetchi (2002, pp. 34–35).

irrespective of age is entitled to ask for a attendant to be present. Such requests should be acceded to whenever possible".[5]

These are the instructions for Kuwaiti doctors: during a vaginal examination, the patient should remain covered as far as possible by a sheet; if the patient is being examined by a male doctor, a nurse or other woman must be present. When the patient is being examined by two male doctors, the one not examining must remain at the level of the patient's head. During labour[6] the pregnant woman may instinctively grip the doctor's hand or arm to obtain psychological comfort. However, the doctor must not encourage such behaviour and should ideally be replaced by a nurse. Relations with female patients must be characterised by kindness, care must be taken to not encourage relations especially outside the hospital setting.

16.3 Some Problematic Areas

According to international bioethics standards (lay, secular, or religious), protecting vulnerable women before they undergo clinical tests, diagnosis and therapy is based on a recognition of their fundamental rights and on the principle of male-female equality. In health care, this means autonomous choice by women without mediators or conditions set by male tutors (*in primis*), the family, the clan or society. In fact, since the origins of humanity women in traditional and patriarchal societies and cultures have universally been subject to social conventions and pressures (i.e., "controlled") more than men. This is also the case of health care subjects whose social role historically has been limited to that of a mother and a housewife[7]. In the Koran, man and woman are considered equal, but different by nature.

The primary task of man is to sustain the family economically; the primary role of the woman is, on the other hand, to raise the children. These convictions affect aid for patients. The (generally elderly) man makes the decisions while the woman bears the burden of care.[8]

The greatest resistance to juridical innovations (*bida'*) on the question of equality has historically been put up by family and inheritance laws since they are directly inspired by the Koran and the *Sunna,* consequently making them more difficult to change. Nevertheless, the influence of Western models on modern Muslim states has begun to "modify" these bastions[9] which directly concern the role and the rights of women. For example, they have affected multiple aspects of traditional marriage contracts and classical rules on the dissolution of marriage; the power granted to the husband and the marital tutor to correct and constrict the woman has become more

[5] See the Code in www.pmdc.org.pk/ethics.htm.

[6] Abu Zikri (1976).

[7] Resolution No. 38 (13/4). Council of the Islamic Fiqh Academy, Fourth Session, Jeddah, 1408 Jamada Tani H. (February 1988); Recommendations, FIRST e), in FIQH ACADEMY, Resolutions and Recommendations of the Council of the Islamic Fiqh Academy 1985–2000, Jeddah, IRTI-IDB, 2000.

[8] Moazam (2000), Lawrence and Rozmus (2001).

[9] Aluffi Beck-Peccoz (1990, 1997).

limited; the recourse to repudiation and polygamy has been limited by several legislations but without abolishing them, as they are regulated by the Koran, with some exceptions such as in Tunisian legislation; the law on minimum age for marriage has been modified in order to fight pre-arranged and early marriages; and new rules aimed at regulating full paternal authority (*wilayah*) and maternal custody (*hadanah*) of the children in various countries tend (timidly) to protect the minor and the mother in the case of dissolution of the marriage contract.

The prohibition of Muslim women to marry non-Muslims remains in all Muslim countries as a reason for the nullity of a marriage. In 1979, the United Nations Convention on the Elimination of all Forms of Discrimination against Women asked its member states to implement a very long set of measures to change or abrogate every law, practice or habit that represented a discrimination against women. Art. 9, but above all articles 15 and 16, deal with family and marital relations. Many Islamic states strongly opposed these ratifications (especially art. 16 paragraphs 1c, d, f, g) stating that only *Shari'a* rules could regulate the institution of marriage and all relations between spouses.

Despite these obstacles, of which only some are mentioned here, bioethics plays a part in encouraging a new male-female equality. For instance, we see this in the promotion of female patient's' full right to information without third parties' interference. We also see the possibility and capacity of women to make autonomous decisions regarding diagnosis and therapy which concern them directly.

Beyond these declarations, however, the real problem remains in limbo: the full and sovereign individual autonomy of the woman only finds a valid and coherent juridical framework in the contemporary democratic Western countries. This anthropological-juridical model considers every person, *man or woman, indistinctly,* as a bearer of inalienable fundamental rights. This priority considers the rights and values of every other institution, social relationship (family, social group, clan, state) or religious-legal principle as secondary. On the other hand, the Muslim woman and her family tend to remain a single unit and the Muslim family continues to be essentially patrilineal and patriarchal, though with increasing variations. The female Muslim patient acts in specific family and social contexts and her values do not align with those of strongly liberal and individualistic tendencies.

These differences have repercussions on the questions of patient decision-making, autonomy and free access of information. Islamic juridical-ethical and deontological claims should take this into account without inhibitions from international documents.

16.4 The Physically and Mentally Handicapped

During the golden age of Islam the most important hospital in Egypt, the Mansuri Hospital (completed in 1284 AD), described its activity in one of its founding documents.[10] It concerns the treatment of the patients, irrespective of their gender, eco-

[10] Rahman (1989, p. 70).

nomic conditions, place of residence, race and religion. Major or minor disorders were treated, including both physical and mental problems. Greatest attention was devoted to those who suffered from the loss of their minds and honour. Individuals as well as entire groups were admitted until they were healed and all the costs were borne by the hospital.

The insistence that mental illnesses are connected to physical disease shows the importance of psychosomatic medicine in classical Islam.

Early Islamic jurisprudence discussed how to protect the handicapped from financial risks, how and whether to recognize the signature of deaf people, and to what kind of rights mentally retarded males were entitled. Due to the attention Egyptian experts traditionally devoted to these topics, the majority of Arabic mental health professionals today are Egyptians, while Egypt seems to remain the most advanced in this field among Arab countries.[11]

In the Islamic context, the value of the human being is closely related to his spiritual development. The Prophet Muhammad said in *Sahih Muslim*, Book 032, number 6220: "Verily Allah does not look to your bodies nor to your faces but He looks to your hearts."[12] The Koran does not stigmatize people affected by disabilities even though only a few references are directly devoted to them. Islam considers disability neither a curse nor a blessing, but as morally neutral[13]. Despite this assessment, in Islamic societies disabilities are very often considered as punishments from God, regardless of their specific causes, physical, mental or otherwise.

Believers can respond to disabilities in two main ways: passively or actively. In turn, these alternatives can be enlightened by two different explanations associated with: (1) religious-magic explanations and (2) biomedical explanations.

1) Thanks to evidence from religious texts, many Muslims believe in the bad and evil effects of whispering (*Waswaas*), magic (*Sihr*), the evil eye (*Al-'Ayn or Nazr*), envy (*Hasad*) and *jinn* (spirit) possession.[14] Some Muslims attribute symptoms of obsessive-compulsive disorder to the action of *jinn*. Others attribute their mental or life problems to magic performed against them by others based on Koran 2:102. Epileptic fits may represent the physical expression of *jinn* possession, even though epilepsy is mostly organic in nature. Among other symptoms attributed to *jinn* are unusual strength, catatonic symptoms, voice changes, anaesthesia to pain and psychosomatic pains[15]. Magic may influence or determine physical and/or psychological symptoms.
Patients can seek protection in all these cases by having recourse to faith, prayers, acts of worship (pilgrimages, fasting, alms), attending a shrine and similar places. It is very important to note that patients may be reluctant to discuss these beliefs with non-Muslim clinicians.
2) When the illness is explained in biomedical terms and references, the patient will attend a medical centre.

[11] Endrawes et al. (2007).

[12] University of Southern California, www.cmje.org/religious-texts/hadith/Muslim.

[13] Hasnain et al. (2008).

[14] Utz (2012, pp. 15–31); Al-Habeeb (2010).

[15] Utz (2012, pp. 19–22)

At times, both of the above-mentioned levels of assistance are sought. It is easy to encounter sociological, religious and/or scientific interpretations which maintain that certain psychiatric and psychological pathology are caused by direct or indirect lack of faith. As a consequence, all sinful societies would easily be affected by some mental illnesses. This does not mean that everyone affected by mental illnesses is detached from Allah. But it is a widespread opinion that this detachment increases the likelihood of problems and the influence of the *jinn*. (Koran 20:124 "But whosoever turns away from My Message, verily for him is a life narrowed down, and We shall raise him up blind on the Day of Judgment").

Those who turn away from Allah may experience depression, anxiety, grief and stressful events. According to the interpretation by Ibn Kathīr (1301–1373 AD), Koran 20:124 means that this person's life will be difficult. Even if he appears to be happy and comfortable, he will not be happy inside as his heart does not have the true guide. He will be agitated, confused and doubtful.[16]

Several diseases could be treated or relieved by praying and reciting (*ruqyah*) the Koran (e.g., Koran 17:82).[17] In any case, it is a very widespread opinion that *qadar* (divine will and predestination) should be always considered in the treatment since drugs, therapy and doctors will succeed in curing the patients only thanks to God's will.

Many Muslims seem to believe more in the efficacy of religion than professional help.[18] When a person experiences depression or other pathologies, his personal endeavour to change this negative cognition may be considered a *jihad*[19]. In fact the Prophet distinguished between "great *jihad*" and "minor *jihad*." The first refers to the "struggle against one's desires", bad wishes and deeds, whereas the second means struggles outside our body, including the classical war. The first meaning is by far the most relevant. In the patients' struggle against psychological and psychiatric illnesses, physicians could make reference to *jihad*.

Muslims believe the family to be an institution of divine origin and the source of happiness, even though it can be a source of intimidation and pressure for its members. The individual-family dialectic in the Islamic worlds appears fundamental in bioethics:

> In many Muslim countries (Egypt, Pakistan, Turkey, etc.) the desires of patients are often subordinate to those of the family and/or of the social group to which the patient belongs. In developed and secular Western societies, the autonomy of the patient represents the fundamental element of medical ethics. On the contrary, in the majority of the world's population, the family (or other intermediate social categories such as clans, social groups etc.) still often plays a decisive role in the decision-making process on the health of one of its members, especially for the purpose of offering psychological protection. In traditional contexts, the individual member is bound by a strong internal hierarchy and strong social bonds. The elder members have a central role in decisions, whilst women are responsible

[16] Utz (2012, p. 18).

[17] Sooki et al. (2011).

[18] Osman M.A. and F. Aboul-Fotouh (2012).

[19] Haque and Kamil (2012).

> for caring for and looking after the patient. In Muslim contexts, reverence for the figure of the doctor, deemed an instrument of divine compassion, can also be added.
> In other words, personal identity can take secondary position with respect to the collective family identity… Alternatively, the decision can easily be delegated to the most authoritative figure in the context in which it is taken, e.g. father, doctor, etc. When the decision-making power of the family is strong, the interests and desires of the "weak" members may be overlooked. For example, when the inequality of power between men and women is strong, the subordination of the woman to her husband's wishes or to those of another male, is very likely. In very poor and/or disadvantaged contexts, the survival of a single family member may be sacrificed due to economic demands, excessively expensive treatment etc. for the survival of the group.[20]

In both traditional Muslim contexts and otherwise, the patient tends spontaneously to delegate the faculty of deciding on the best course of treatment to relatives especially when the condition is serious. At the same time, there is a tendency for Muslim families to instinctively take on the wishes (even those that are not expressed) of the patient. In practice, the patient's decisions in consultation with the doctor often take the family into consideration. This could be a limit on the patient's autonomy. In the prevalently psychological-psychiatric perspective, "for many Muslims, the self is not often differentiated from the family identity, thus self-concept and self-esteem will have collective meanings, depending on the family's reputation and approval and reflections of the family's identity."[21]

It is understandable how the fundamental role played by individuals in the new codes of law (civil, criminal, etc.) with a Western-international stamp does not reflect at all the age-old mentality, customs and habits still present in many areas of the Islamic world and elsewhere.

One interesting distinction is between acquired and lifelong disabilities[22]. When faced with acute or acquired illnesses, many Muslims have faith in a cure. At the same time they may not view a chronic condition as an illness, despite its disabling features. In Afghanistan and Pakistan, more resources are placed to support those with traumatic disabilities than congenital ones. The first elicited more feelings of respect whereas those with congenital disabilities were spoken of paternalistically.[23]

According to Islamic Law, the Muslim community (*Umma*) takes priority over all other human communities (*dhimmis,* atheists or polytheists). Within the *Umma*, community claims priority over the family while, in turn, the family has priority over the individual, above all in the cases of disabilities. The family is the basic social institution which pitches in when there is absence or shortage of public care of the physically or mentally handicapped and the elderly. The man-woman or boy-girl difference generally remains relevant in many traditional societies including in the treatment given to handicapped females.

[20] Atighetchi (2007, pp. 49–50, 274–283).

[21] Daneshpour (2012).

[22] Hasnain et al. (2008, pp 38–39).

[23] Hasnain et al. (2008, pp 38–39).

16.5 Some Problematic Data

Contemporary "Western" psychiatry and psychology, thanks to their conceptual categories, terminology, health and therapeutic organizations, normative models, etc. have influenced the development of psychology and psychiatry in other cultures. In this field, the problem of acculturation of these conceptions to the detriment of autochthonous origins is clear.

There is widespread request that the psychologist-psychiatrist treating a Muslim should know the basics of Islam to start an effective course of therapy[24]. The risk is that the "true faith" creates a "particular" or a "*sui generis*" patient. In this case, non-Muslim psychotherapists and psychiatrists are unsuited to understand the characteristics or secrets of the believer's mindset. A specificity-inimitability of the psychological-mental mechanisms of the Muslim patient profile is intrinsically characterized by faith.

Religious belonging is all the more significant when the family does not accept scientific diagnosis or clinical references and turns instead to religion in managing mental illnesses (e.g. schizophrenia).[25] The important "Islam factor" is further shown by development of instruments to measure Islamic religiosity (practices and beliefs) in mental health research, such as Religiosity of Islam Scale (RoIS) or Psychological Measure of Islamic Religiousness (PMIR), etc.[26] The question arises as to whether fundamental psychological mechanisms are universal or not. By inverting the perspective, can a Muslim psychologist and/or psychiatrist treat a secular or atheist patient effectively?

The Diagnostic and Statistical Manual of Mental Disorders (DSM) is the "Summa" of Western psychiatry and represents the most important systematic description of human psychopathologies. Regardless of culture or traditions, it is almost impossible today to study these pathologies scientifically without making reference to these texts. Western psychological/psychiatric currents and schools are generally non-religious, not anti-religious. In recent years, there has been a growing interest in the relation between religiosity/spirituality and mental health in the therapy of patients with strong religious sense. Research indicates that religious Muslims who practise their faith "experience higher levels of happiness, well-being, life satisfaction and marital satisfaction and have a reduced likelihood of depression, anxiety, death anxiety, antisocial behaviour and suicide."[27]

This observation is also valid for the faithful of other religions. Therefore, in some cases, religious faith seems to create a "different" patient or one less exposed to pathological states such as depression and anxiety.

Obviously, there is no guarantee that a therapy will be successful and that faith will bring mental and/or physical benefits since only God (the True Healer) decides

[24] Dharamsi and Maynard (2012).

[25] Osman M. Ali and F. Aboul-Fotouh (2012, p. 41).

[26] Jana-Masri and Priester (2007); Abu Raiya et al. (2007).

[27] Utz (2012, p. 15).

the outcome. There are effectively proximate and ultimate causes for illness. Biochemical modifications should be considered proximate causes, whereas Allah's Will is the ultimate cause. While health professional deals with proximate causes, faith explains the ultimate cause. Both motivations could act together, in alliance.[28]

In the Islamic perspective, the human psyche is not a neutral terrain governed by neutral rules of treatment where reference to the Creator can be disregarded. On the contrary, real treatment must be anchored in the origin of each healing and illness. This is why the references to God and the *ahadith* represent an irreplaceable foundation.

A neuroethical question follows: to what extent can religious faith modify or characterize—in ways that can be scientifically observed—the mind, the psyche, psychic mechanisms or psychology of man? Prayer is often considered a psychotherapeutically useful practice in several psychological disorders. However, the precise effects of prayer at the neurological and neuroethical levels are still little known. Is it a self-illusion (placebo effect and the like)? Can prayer and faith be an exclusive source of benefits that is technically ascertainable? If the latter is true, are there prayers that are universally effective and indifferent to the person's faith, or are some prayers more effective than others?

Due to the aforementioned characteristics, Islamic psychology should be defined as psycho-theology, "theistic psychotherapy" or "religious psychotherapy" since reference to the Word of God and examples (words, actions and silences) of the Prophet Muhammad should form the structure for treating Muslim patients.

Many therapeutic approaches are effective with Muslims. One useful therapy is the "solution-focused model" aimed mainly at finding practical solutions to problems and limiting self-disclosure by the patient. In addition, this model allows integrating religious concepts when working with religious patients. Another useful therapy is the so-called "metaphor therapy" which integrates of Koranic metaphors "as a healing approach." These approaches "emphasize spirituality, communal relationships, benevolence and clarity of consciousness" while incorporating "prayer, mediation and belief." "This may be a suitable fit for Muslim clients in contrast to Western therapies that are secular and emphasize individualism and independence" [29].

For most Muslims, psychoanalytical/psychodynamic approaches are not suitable. Cognitive-behavioural techniques are generally the most effective, but "mental health practitioners working with Muslims should be flexible and may need to accommodate modifications to their theories or interventions" [30].

Most Western psychopathology theories (e.g. cognitive-behavioural and person-centred models) focus on the individual and his or her interiority. They are inadequate when they ignore the internal dynamics of the family. On the other hand, the Family System Theory (FST) is a practical model for Muslims as it represents an approach that is attentive to the roles of each family member of different genera-

[28] Osman M. A. and F. Aboul-Fotouh (2012).

[29] Amer and Jalal (2012, pp. 111–112).

[30] Amer and Jalal (2012, pp. 113–114).

tions.[31] It interprets the family as a social unit, which is not simply the sum of its parts. The action of each member influences all the other members of the family. FST identifies the family as the centre of emotive development, behavioural models, values, and bonds of loyalty, etc. of each individual member.

The family represents a "small society" with a high degree of social responsibility. As a consequence, each member is responsible for the healthy functioning of all the others. The appearance of psychological or psychiatric symptoms in a single member may reveal the presence of dysfunctions in the "family system" [32].

16.6 The Elderly

I will focus my attention on three aspects related to the vulnerability of the elderly: namely palliative care, hospices and the dying process. First, however, an introduction on the situation of the elderly in some Islamic societies is in order.

Until recently, the lives of children could be almost identical to those of their parents. The slow, practically imperceptible, cultural changes in the distant past looked upon the elderly as custodians of knowledge, traditions, techniques for survival, and social and religious values. For peoples without writing, the elderly represented "living cultural computers" of society. The community's respect for the elderly was fundamental to their survival and contributed to one of the foundations of the patriarchal system.

With modernity, there has been an impetuous acceleration of cultural changes in the ways of life. The scientific revolution has swept away traditional social models based on the elderly as custodians of traditional knowledge and values. Computers that are efficient, fast, and have exceptional performances that can be constantly updated are preferred to the wisdom of the elderly. With accentuated modernization, the elderly risk becoming an icon of something useless and irremediably outdated. Modernization goes hand in hand with individualization. Thus, reports of a reduction in the protection of the elderly are present in many Islamic areas undergoing modernization.

In the West, the elderly no longer have active and useful roles that were present in societies of centuries ago. At the same time, the lengthening of the average lifespan is coupled with declining cerebral and mental performances. The elderly retiree risks "surviving" for decades with mental performances that are less competitive in societies that require, paradoxically, greater capacities of cultural adaptation and technical-scientific skills. In the West, we are seeing an unprecedented ageing of the population with problems which have never been faced before.

The ageing of the population in Morocco provides a sample of the Islamic world which, in spite of being demographically young, progressively encounters an ageing population. In 1982, Morocco was one of the six countries that declared old

[31] Daneshpour (2012, p. 122).

[32] Daneshpour (2012, p. 122).

age not to be a priority for the State but, according to traditional practice, a family concern in which the elderly are cared for by their descendants. Other countries have recognized, at least theoretically, that "family management" of the elderly is no longer sufficient in the contemporary world. At a press conference in 1986, when the World Health Organization (WHO) recommended a plan of action for the elderly that included the need to open homes for the aged, King Hassan II gave the following response: "the day when we open an old people's home in Morocco, our society will be about to disappear" [33]. He added that if someone were to build one in Morocco, the King himself will set fire to it.

At the World Assembly on Old Age held at Madrid from 8–12 April, 2002, the Moroccan *Rapport national sur le vieillissement* states that,."..ageing is not at the present moment… an actual social phenomenon, taking into account objective considerations that provide solid foundations to preserve human dignity, in particular social, cultural, 'of civilization' and religious ones."

In the North African collective imagination and in mass media, the elderly still represents the ideal icon of protectors of national, social, religious values and identities; values and identities that modernity tends to relativize and overturn[34].

Koran 17:23–24 associates the divine order of men worshiping a single God with men's duty to respect their parents. As a result of this close link, sociologist Rahma Bourquia explains that "the blessing of the parents… erected as a value by religion, beliefs and the collective imagination, acts as a principle that preserves the cohesion of the family through time. Parental blessing accompanies the individual throughout his life.... Being banned by parents is equivalent to being banned by God, the family and society."[35]

Solidarity with the elderly is often perceived as an identity value, especially in northern Africa. It is regarded as a confirmation of the persistent social appreciation of non-economic values. In reference to Tunisia, M. Nasraoui summarizes: "Arab-Muslim society has a favourable attitude towards the past, is ambivalent towards the present and worries about the future; old age is naturally idealized" .[36]

The distinction and difference between genders remain important in old age. Indeed Mokhtar El Harras states: "The elderly man is better seen and is more respected than the elderly woman" .[37]

In Muslim countries, home care and the presence of extended families represent the best (or only) solution for providing psychological and material support for the patients. In addition, thanks to the strong family unit and the influence of religion, the elderly and incurable patients generally die at home. In many areas, the family still acts as a protective shield from the sorrows and psychological pains of the patient. This protection often requires not revealing to the patient his real condition.

---

[33] Jacquet (2009, pp. 190–192).

[34] Jacquet (2009, p. 140).

[35] Jacquet (2009, pp. 135–136).

[36] Jacquet (2009, p. 141).

[37] Jacquet (2009, p. 141).

It is paradoxical that lengthening of lifespan is accompanied with longer periods of pain despite the growing availability of various types of painkillers. Given that today, death is no longer an event but more and more a long and painful process, religious, spiritual and ritual assistance should be provided. Meanwhile, we should avoid giving alibis to not intervene in these painful situations using appropriate drugs. According to some recent data, in the period of 1999–2007, consumption of morphine for relief of cancer pain in the USA was 10 times that of Israel, 27 times that of Cyprus, 38 times that of Jordan, 69 times that of Lebanon, 150 times that of Saudi Arabia, 447 times that of Turkey and 663 times that of Egypt.[38] Even though this is a singular datum, the general trend is impressive: elderly patients dying in these contexts undergo very painful experiences.

Today's terminally ill patient needs palliative care.[39] Recourse to hospices or similar structures in the Middle East and North Africa is hindered by multiple reasons. In the first place, they are of an economic and social order since complex and expensive infrastructures are needed. Then, there is the problem of the inefficiency of nurses and the poor qualification of health personnel with low salaries. Lastly, there is a general preference for home care.

Other difficulties are similar to those in many developed countries and can be summarised as follow[40]: emphasis on "cure" when recovery is no longer possible, scarcity of doctors interested in this type of approach, absence of adequate methods of pain control, reluctance to discuss the problems of dying and death, general refusal to inform the patient of the diagnosis of cancer and the fatal prognosis.[41] In many Middle Eastern and Islamic countries, the population tends to identify a cancer diagnosis with a death sentence. Therefore, doctors are reluctant to reveal it.

In 1999, the European School of Oncology sponsored a symposium at the King Faisal Specialist Hospital in Saudi Arabia entitled "The Modern Management of Advanced Cancer: How to help your patients".[42] This symposium analysed the availability and distribution of opioids in the Kingdom. This Saudi debate represents the prevailing situation in many Muslim countries. The main problems identified were: (1) General lack of awareness of the importance of palliative care and scarce support from the community. (2) Lack of preparation on the issue among doctors, nurses, patients and relatives.

As a result, there are mistaken ideas and fears of interfering with the patient's religious duties. There is a great lack of experience about the "analgesic ladder." (3) There is a general fear that use of opioid analgesics goes against religious principles. The final recommendation was to instruct religious authorities on the neces-

---

[38] Silbemann (2010).

[39] Atighetchi (2007 p. 272).

[40] Atighetchi (2007 p. 273).

[41] There is an enormous bibliography on these subjects. I mention only: Babgi AA (2010). Pain Coping Behaviors of Saudi Patients suffering from Advanced Cancer: A Revisited Experience. Asian Pacific J. Of Cancer Prevention 2,MECC Supplement: 103–6; Daher M. Pain Relief is a Human Right, ibidem: 97–101.

[42] Isbister and Bonifant (2001).

sities and methods of care for the terminally ill and to include them in the palliative care staff. Another suggestion was a mass media campaign of education to promote a national policy on coping with pain in the terminally ill patient.

It is interesting to note that while developing countries cover the large majority of the world population today, their opioid use for pain relief takes up only 8% of the world consumption.

Palliative care is strongly associated with culture. Cultural factors can influence different dimensions of pain as well as the quality of care. The Islamic world seems to experience difficulty in adapting its heritage to effectively meet the challenges of physical and psychological pain.[43] The solution remains mainly linked to the classical religious dimension[44]: family psychological care, faith, prayer, and acts of worship.

There are greater difficulties for the elderly because classical networks of social protection have diminished while at the same time these societies are late in updating their knowledge and technology to fight against physical and mental pain.

References

Abu Raiya, H., and K. I. Pargament, et al. 2007. Lessons learned and challenges faced in developing the psychological measure of Islamic religiousness. *Journal of Muslim Mental Health* 2 (2): 133–154.

Abu Zikri, A. 1976. Ethics in obstetric practice. *Journal of the Kuwait Medical Association* 10:191–196.

Al-Habeeb, T. A. 2010. A pilot study of faith healers' view on evil eye, jinn possession and magic in the Kingdom of Saudi Arabia. *Saudi Society of Family and Community Medicine* 10 (3): 31–38.

Aluffi Beck-Peccoz, R. 1990. *La modernizzazione del diritto di famiglia nei paesi arabi*. Milan: Giuffrè Editore.

Aluffi Beck-Peccoz, R. (ed.) 1997. *Le Leggi del diritto di famiglia negli stati arabi del Nord-Africa*. Turin: Edizioni della Fondazione Giovanni Agnelli.

Amer, M. M, and B. Jalal 2012. Individual psychotherapy/counseling. In *Counseling Muslims*, ed. S. Ahmed and M. M. Amer, 87–117. London: Routledge.

Atighetchi, D. 2002. *Islam, Musulmani e Bioetica*. Rome: Armando Editore.

Atighetchi, D. 2007. The Koran and modern science. In Atighetchi D. *Islamic Bioethics: Problems and Perspectives*, 327–351. New York: Springer.

Atighetchi, D. 2009. *Islam e Bioetica*, 217–220. Rome: Armando Editore.

Castro, F. 1981. Sistema sciaraitico, "siyāsa šar†iyya" e modelli normativi europei nel processo di formazione degli ordinamenti giuridici dei paesi del Vicino Oriente. In *Il mondo islamico tra interazione e acculturazione*, ed. A. Bausani and S. Amoretti, 165–202. Rome: Università di Roma - Istituto Studi Islamici.

Castro , F. 1985. La codificazione del diritto privato negli Stati arabi contemporanei. Rivista di Diritto Civile I: 377–447.

Castro, F. 1990. *Diritto Musulmano, in Digesto delle discipline privatistiche 6*, 1–66, 14–15. Torino: UTET.

[43] Schultz et al. (2012).

[44] Sooki et al. (2011).

Daneshpour, M. 2012. Family systems therapy and post modern approaches. In: Counseling Muslims, ed. S. Ahmed and M.M. Amer, 119–134, 120–121.

Dharamsi, S., and A. Maynard 2012. Islamic-based interventions. In: *Counseling Muslims*, ed. S. Ahmed and M. M. Amer, 135–160. London: Routledge.

Endrawes, G., L. O'Brien, and L. Wilkes 2007. Mental illness and Egyptian families. *International Journal of Mental Health Nursing* 16:178–187.

Haque, N., and A. Kamil 2012. Islam, Muslims and mental health. In *Counseling Muslims*, ed. S. Ahmed and M. M. Amer, 3–14, 11. London: Routledge.

Hasnain, R., L. Cohon Shaikh, and H. Shanawani 2008. *Disability and the Muslim perspective: An introduction for rehabilitation and health care providers*, 31. Buffalo (New York): University of Buffalo, CIRRIE.

International Human Rights. 1993. p. 325. Wien.

Isbister, W. H., and J. Bonifant 2001. Implementation of the World Health Organization. Analgesic ladder in Saudi Arabia. *Palliative Medicine* 15:135–140.

Jacquet, I. 2009. *La vieillesse au Maroc*. Louvain-la-Neuve: Academia Bruylant.

Jana-Masri, A., and P. E. Priester 2007. The development and validation of a Qur'an-based instrument to assess Islamic religiosity: The religiosity of Islam scale. *Journal of Muslim Mental Health* 2 (2): 177–188.

Lawrence, P., and C. Rozmus 2001. Culturally sensitive care of the Muslim patient. *Journal of Transcultural Nursing* 12 (3): 228–233, 230.

Moazam, F. 2000. Families, patients and physicians in medical decision making: A Pakistani perspective. *Hasting Center Report* 30 (6): 28–37, 30.

Osman, M. Ali, and F. Aboul-Fotouh 2012. Traditional mental health coping and help-seeking. In: *Counseling Muslims*, ed. S. Ahmed and M. M. Amer, 33–47, 35. London: Routledge.

Rahman, F. 1989. *Health and medicine in the Islamic tradition*, 70. New York: Crossroad.

Schultz, M., and K. Baddarni, et al. 2012. Reflections on palliative care from the Jewish and Islamic Tradition. *Evidence-Based Complementary and Alternative Medicine* 3. doi:10.1155/2012/693092.

Silbemann, M. 2010. Opioids in Middle Eastern populations. *Asian Pacific Journal of Cancer Prevention* 11.

Sooki, Z., and K. Sharifi, et al. 2011. Role of Quran recitation in mental health of the elderly. *Quarterly of Quran & Medicine* 1 (1): 17–23.

The Holy Koran. (trans: Yusuf AA). 1983. Maryland: Amana Corporation.

Utz, A. 2012. Conceptualizations of mental health, illness and healing. In: *Counseling Muslims*, ed. S. Ahmed and M. M. Amer, 15–31, 21. London: Routledge.

Chapter 17
Jewish Perspective on Vulnerable Groups: The Elderly and the Disabled

Jonathan Halevy and Adina Halevy

17.1 Introduction

According to Jewish tradition, G-d created all human beings in their different shapes, characters and forms. G-d sees the human being as his partner in creation and in molding and advancing the world. As we are all partners of G-d but none of us is like G-d—"yet Thou hast made him a little lower than the angels" (Psalm 8:5)—all of us are, to a certain extent, disabled.

The divine command, recorded in the Torah in its perfect and harmonious fashion, is directed to us humans—as individuals and as a society. The human being, trapped within the confines of the human condition has, on the one hand, a body, and on the other, is endowed with the divine attribute of partnership in Creation and advancement of the world. He is constantly confronted with the need to adapt the spirit of that divine command to the imperfections his mundane reality. The question of whether these imperfections—sickness, disability, mortality, ugliness and the like—are the handiwork of G-d or the responsibility of man is illustrated in the Talmud (Ta'anit 20). In a story, Rabbi Elazar, the son of Rabbi Shimon, was returning home riding his donkey after an extended period of Torah study and contemplation, when:

> …he encountered a man who was very ugly. Rabbi Elazar greeted him, but the man did not respond. Rabbi Elazar then said to him: "What an ugly man! Are all the people in your town as ugly as you?" And the man replied: "I don't know. Go and tell the artist who made me: How ugly is this vessel that you have created!" Realizing that he had sinned, Rabbi Elazar dismounted, prostrated himself before the man and said: "I have responded to your plea—forgive me!" But the man replied: "I will not forgive you until you tell the artist who made me: How ugly is this vessel that you have created!"

J. Halevy (✉)
Shaare Zedek Medical Center, Jerusalem, Israel
e-mail: halevy@szmc.org.il

A. Halevy
Jerusalem, Israel

J. Tham et al. (eds.), *Religious Perspectives on Human Vulnerability in Bioethics*, Advancing Global Bioethics 2, DOI 10.1007/978-94-017-8736-9_17,

The gap between the harmonious light of the Torah and the encounter with ugliness comes as a shock to Rabbi Elazar. While he was engaged in the study of Torah, he was deeply immersed in his aspiration to an angel-like perfection that clashed rudely with the ugliness of earthly reality he suddenly encountered—a reality that makes one realize that G-d created man in all his facets and attributes. The way in which the divine command is to be realized here, in our human and physical world, is the amalgamation of several elements: the handicapped individual's objective condition; his subjective condition; the phenomenology of his condition; the socio-cultural milieu of the period; and the use of science and technology as a means of overcoming the limitations of a given situation.

Throughout history, there have been many discussions and deliberations in Jewish sources as to the proper place of the physically and mentally handicapped in society, and the proper attitude of the society at large, as well as of individual members in particular towards them. Jewish Law *(Halachah)* demonstrates a dynamic approach, attentive to a variety of voices, in its ongoing quest for ways and means to give expression to the spirit of the divine command in keeping with a changing mundane reality. Our discomfort when we face the disabled stems from the primordial archaic fear that we too may fall sick or become disabled, and from the always existing threat to our health (Sontag 1978). Vulnerability and disability of the individual challenge all components of society: nuclear family, extended family, the community at large and various state agencies.

Life is always vulnerable to what may affect its quality and normal flow. Nonetheless, there is an essential difference between vulnerability due to old age perceived as part of the normal life cycle of biological-developmental, and physical, sensorial or mental disabilities that are defined as diseases and may stigmatize the persons afflicted with them. Since Jewish tradition does not relate to "handicap" as a general term (Shoham-Steiner 2008), we shall discuss three examples of the fact that imperfection is an inseparable part of human existence: old age, insanity and sensorial disability.

17.2 The Elderly

A typical definition of old age is that given by Wikipedia: "Ages nearing or surpassing the average life span of human beings and thus the end of the human life cycle" (Wikipedia entry "Elderly.") In Jewish law, the definition of old age varies according to the situation. In certain circumstances, old age is defined purely according to chronological age but, even then, the age varies in regard to different legal situations. Old age is also defined according to physiological characteristics in specific circumstances (Steinberg 2003).

The word "*Zaken*" (old/elderly) in the Bible refers to chronological age. However, the term is used in the Talmud as a metaphor for an important or esteemed person, such as a great Torah scholar, even if he is young in years. This interpretation follows the acrostic of the Hebrew word ZAKEN, whose Hebrew letters spell out the phrase "one who has acquired wisdom" (*Zeh Kanah Chochma*, Kiddushin 32a). The word "*Zaken*" is also used metaphorically to denote an honored leader.

These various metaphorical uses of the word "elderly" indicate the esteem and respect that Jewish thought bestows upon seniors as a result of the wisdom and experience acquired with advancing age. The basic Jewish approach is to accommodate the elderly within the family environment. In the Bible and Talmud, the elderly are not included in the groups of underprivileged people such as the poor, the proselyte, the orphan and the widow who require special considerations of charity and compassion because of their disadvantaged status. Similarly, there is no mention in the Bible and Talmud of institutions and/or organizations specifically designated to helping the elderly. Thus, we learn that the integration of the elderly into family life and the care of the elderly by the family are the normal and accepted practice (Steinberg 2003).

In Jewish sources in general, an elderly person is regarded as a fully active individual, although his limitations, weaknesses and problems are touched upon extensively in the writings of the rabbinic sages. Old age is viewed positively insofar as it is the stage of life associated with accumulated life experience and knowledge, and with considerable free time to study the Torah and fulfill G-d's commandments. Thus, old age is considered to be a blessing (Isaiah 65:20, Zechariah 8:4, Psalms 92:15, Proverbs 3:16) although, excessive longevity is not (Tractate Berakhot 7a). Given the reverence due to the elderly, shaming an old man is condemned as a sign of corruption (Deuteronomy 28:50, Isaiah 3:5). One of the indications of the deterioration of ethical behavior after the destruction of the Temple was that youths put old men to shame and the old stood in the presence of the young (Sotah 4ab, Sanhedrin 97a). One who learns from the old is compared to one who eats ripe grapes and drinks old wine (Avot 4:20) implying that the teachings of the old are mature, correct and accepted. In the Talmudic academy, none is more fitting than an old man (Chagigah 14a).

"Thou shalt rise up before the hoary head and honor the face of the old man" is one of the 613 Biblical commandments (Leviticus 19:32). He who rises before an old man is obligated to rise to his full height and to remain standing as long as the old man is standing (Shevuot 32b). Therefore, when riding on public transport, one should offer one's seat to the elderly. The definition of old age for the precept of rising is seventy years (Avot 5:24). Jewish sources give much respect to the elderly, even when signs of cognitive dysfunction begin to appear. An elderly person who has forgotten his learning through no fault of his own is treated like a "Holy Ark," (Yerusahlmi Moed Katan 3:1) the place where both the tablets containing the Ten Commandments and the fragments of the tablets were placed.

An old man who becomes senile is not deposed from his position of honor (Exodus Rabbah 5:12; Leviticus Rabbah 11:8). In many places in the Bible we find that G-d honored the elderly (Reishit Chochma, Shaare Hayirah 15:10). Certainly, if the Creator of the World honors the elderly, how much more are human beings obligated to do so. Another source states that whoever does not respect the elderly will not live long (Bava Batra 142b). At the same time, Jewish sources recognize the limitations of old age and the fact that if a person did not acquire knowledge and did not concern himself in doing good in his youth, he may become a burden on himself and on others during his old age (Bava Batra 142b).

While Jewish sources mandate that children tend personally to their parents' physical and psychological needs, Maimonides adds one caveat: those who are extremely disturbed by their parents' dementia may arrange to have someone else care for their parents (Maimonides, Mishneh Torah, Ishut 13:14). However, it is gracious for children to open their house to their parents, to feed them and provide for their needs (Maimonides, Mishneh Torah, Ishut 22:3). If an elderly man can neither live alone nor with his children, he should be supported by charitable funds. Refusal of charity by the elderly in such a case is considered a sin (Tur, Shulchan Aruch, Yoreh Deah 255:21).

In the matter of healing or saving a life, no difference is made between a young and an old person, both with regard to the obligation of standard therapy and with regard to priority in treatment (Responsa Igrot Moshe, Choshen Mishpat, Part 2, 75:7). The last decades have witnessed increasing awareness among medical practitioners of the importance of the autonomy of patients, in contrast to the paternalistic attitude that characterized medical practice in the past. This approach, however, has bred its own ethical and practical problems for the geriatric physician, especially in situations of managing life threatening events. In 2005 The Dying Patient Law was enacted in Israel. This law regulates the medical treatment of the dying patient while properly maintaining a balance between the value of sanctity of life, the value of individual autonomous will, and the importance of quality of life.

The fundamental principle of the law is that when medical treatment of the dying patient is determined, the factors exclusively taken into consideration are his medical condition, his will, and the extent of his suffering. In this manner, and only after strict definition of the patient as suffering from an incurable disease with a very limited life expectancy (up to 6 months), the law makes provision to respect the will of the patient and to refrain from life-extending treatment that the patient would not wish to be implemented. Concomitantly, the physician must make sure that every possible act is performed in order to alleviate the pain and suffering of a dying patient through palliative care. The law prohibits active killing, assisted suicide or the termination of continuous medical treatment. In conclusion, Jewish sources teach in favor of an utmost respect for the elderly, and impose upon the children the duty of maintaining the dignity and wellbeing of their elders.

If the children are unable to provide for their elders, this responsibility becomes the duty of society at large. The Jewish approach to the end of life of the elderly gives priority to the sanctity of life while doing the utmost to alleviate pain and suffering, with an option to respect the will of the patient to refrain from life-extending treatment.

17.3 The Mentally Handicapped

The following story in Jewish sources about David, a King-to-be, demonstrates his deliberations about insanity:

> …David said before the Holy One, blessed be He: "All that You have made is beautiful and wisdom is the most beautiful of all…but insanity that You created—what beauty is there in it for You? For example, when a man walks in the market and he drools over his clothes and children run after him and the people make fun of him; this is beautiful before You?"

> The Holy One, blessed be He, said to David: "You complain about the injustice of insanity; by your life you will regret this and you will pray for it until I give it to you."
> Only a short time passed till David found himself with Achish… The Holy One, blessed be He, said to David: "You are going to Achish? Yesterday you killed Goliath and his brother is the head guard of Achish and his blood is still not absorbed [in the ground] and you are going to him with your sword in your hand?"
> David began to plead and pray and said: "Master of the world, answer me at this time." The Holy One, blessed be He, said to him: "What are you asking for?" He said to Him: "The same insanity that you created and I complained about." He said to him: "Didn't I tell you, 'he who despises a word—it will punish him?' Nevertheless I will give it to you."
> David made himself as if insane and changed his whole appearance and he wrote on the walls: Achish the king of Gat owes me a hundred thousand and his wife owes me fifty-hundred thousand. [Now] the daughter of Achish and her mother were insane. They were both screaming and going crazy inside and David was screaming and going crazy outside. Achish said to them [in the court]: "and did you think that I am lacking crazy people?!" At that moment David rejoiced that he was able to feign insanity and from that joy he made this song [the thirty-fourth Psalm]. The Holy One blessed be He said to David: "Is insanity then good?" This is what is written [by David]: "I will bless you at every moment" and also it is said: "everything He made is beautiful in its hour" (Midrash Shocher Tov #34).

The *Gemara* terms a mentally handicapped individual a "*shoteh*" and in response to the question: "Who is a *shoteh*?" (Hagigah 4a) defines the nature of this phenomenon in four different ways:

> He who goes out alone at night;
> He who spends the night in a cemetery;
> He who tears his clothes;
> He who destroys all that is given to him.

The *shoteh* is portrayed as one whom we would understand today as manifesting symptoms typical of the insane, or what in clinical terms is known as the psychotic individual. Psychosis in this sense may be heuristically defined as the state in which the individual lacks the ability to distinguish fantasy from reality. This concept is clinically associated with severe impairment of social and personal functioning and with an inability to perform expected roles. With gross limitations of reality-testing and insight, the psychotic individual incorrectly evaluates perceptions and thoughts and in so doing, makes incorrect inferences about external reality despite evidence to the contrary (Kaplan et al. 1994).

It is clearly evident that the *shoteh* is one who is struggling to meet societal expectations of functioning and who would demonstrate impaired ability to cope with the usual frustrations and routine challenges of life. The norms and values of *Halachah*, the Jewish codex, demonstrate acute sensitivity to the nature of the *shoteh*, thereby displaying awareness of the wide range of human experience in this, as in many other domains. This sensitivity manifests itself not merely towards the mentally handicapped himself—protecting his rights, status and standing in the community—but also towards the rights of the community, which may find itself at times compromised by a psychotic's irrational acts (Strous 2001).

Halachah recognizes that psychotic symptomatology exists on a continuum and that this array of human behavior may express itself in a number of forms; thus it does not lump all phenotypic expressions of psychotic illness together. The *halachic* nuances of the various categories of the *shoteh* have been defined by our Jewish scholar-sages as follows—and we cannot but note the striking parallelism to modern psychological classifications:

1. Completely insane (*shoteh gamur*) best illustrated by the chronic schizophrenic patient, whose prognosis remains poor and whose baseline functioning is seldom regained (Rana, Even ha-Ezer 44).
2. One who moves cyclically in and out of psychosis, like the bipolar patient for whom the prognosis is generally good and who may be considered sane and competent when not in the psychotic state (Ketubbot 20a).
3. One who is insane in only one domain, remaining sane and coherent in all others, as seen in contemporary delusional disorder, or in severe cases of eating disorders. This individual, though remaining psychotic in a specific area, may be judged competent to engage in certain other domains of responsibility and obligation which would otherwise remain off limits to other "*shotim*."

This categorization is very important since it determines the various obligations expected of and permitted to the *shoteh*, sanctioning potential roles, so that his status and rights are secured.

Lacking the critical judgment necessary for basic tasks of daily living and social adaptation, the ability to assess the situation correctly and to act appropriately, the *shoteh* is exempt from the obligation to observe the commandments of the Torah, both positive and negative. Indeed it is forbidden to impose these upon him (Hagiga 2b).

It may be argued that these exemptions from obligation contain an element of discrimination; in actual fact, however, this halachic paradigm demonstrates acute sensitivity. Psychologists today also recognize that adding further stress to an already challenged psychological state of mind may potentially exacerbate the clinical picture, or impede recovery from a currently disabling psychotic illness.

As indication of the *Halachah*'s understanding of the dynamic nature of the mental condition, it should be noted that, while a *shoteh* is exempt from these duties when he is psychotic, once he remits from his state, he again becomes obligated to them (Maimonides Hilchot Hamets U'matsa 6:3).

The Jewish tradition also displays great sensitivity towards society as a whole, thus safeguarding the interest of both the normal and the deviant. Special care is taken in several areas in order to protect the mentally impaired from harm, while simultaneously respecting his rightful place as a member of the community. Here are a few examples of this double-sided concern:

1. In a state of impaired insight and even disconnection from the world, one may be unable to distinguish the nature of cause and effect resulting from his actions. Consequently, such a person is exempt from responsibility for certain damages. He is absolved from accountability if he injures another individual, but, were others to injure him, they would be held accountable for any harm that occurs (Bava Qama 87a).
2. A *shoteh* is unable to take upon himself the responsibility for others' well-being. This is particularly relevant to the role of intermediary or emissary. (For example, blowing the *shofar* (Rosh ha-Shana 29a); baking *matza* (Tur, Orah

Hayyim 460); delivering a bill of divorce, Gittin 9a, 23a). Once again *Halachah* defends against undue exploitation and burdening of the mentally impaired, and protects sane individuals from the vagaries and idiosyncrasies of the psychotic individual who cannot be relied on in case of need and potential dependency.

3. The insane female: It is prohibited to divorce a woman who has become insane, since such abandonment would be cruel, in that she would lose the protection provided by the marriage (Responsa Seridei Eish 3:21). Another example of sensitivity towards the psychotic female is expressed in the opinion that one can be lenient and allow sterilization of a *shotah* (insane female) out of significant concern for her welfare and potential exploitation (Yevamot 112b).
4. The *Halachah* regards psychosis as a life threatening illness, thus the rules limiting actions on days such as *Shabbat* and *Yom Kippur* may be set aside for a patient in a psychotic state since he is considered to be in acute danger (Responsa Tsits Eliezer 8 15:12, no. 5).

Over the past few decades, the field of psychiatry has made great progress. These across- the-board advances have contributed significantly to a better understanding of the brain, both in its normal and in its aberrant functioning and, more importantly, to the successful treatment and relief of many mental disorders, including some of the most severe and previously treatment-resistant psychotic illnesses. This progress may be combined with the *Halachah*'s understanding of the dynamic nature of the mental condition noted above, and its relevance may be absorbed through an updating of the guidelines. In other words, it could be possible to combine the divine spirit and sensitivity expressed in ancient rulings, with the advances of modern psychiatry. For example, the assumption expressed in the Talmud "once insane, always insane" (Gittin 70b), which was true when there was no cure, may no longer apply in light of current medical, clinical and pharmacological knowledge and practice.

This openness and flexibility attuned to new scientific data and clinical knowledge is the outcome of a clear ethical stand which requires that continual attention be paid both to the subjective situation and to changing objective factors.

Although the laws of Israel today are not based on *Halachah*, the spirit of the ancient Jewish laws often prevails in the initiation of our new laws. Among the most important social laws in the modern State of Israel is the "Rehabilitation of the Mentally Disabled in the Community Law" (2000).

Progressive by international standards, this law contributes to the effort to shift the locus of treatment and care from mental institutions to the community. "Rehabilitation" in this respect refers to a process within the framework of the community, directed at the development of the abilities and skills of the mentally disabled person—a process accompanied by medical supervision while upholding the realization of the rights of the mentally disabled in the areas of housing, employment, education and training in professional and leisure skills.

The purpose of this law is to strive for and advance the rehabilitation and integration of the mentally disabled in the community in order to allow them to achieve the maximum degree of functional independence and the highest possible quality of life, while preserving their dignity—a truly modern embodiment of the Halachic spirit.

17.4 The Blind

A blind person is unable to see because of a problem in the eyes, in the optic nerve pathways, or in the brain centers of vision (occipital lobes). A person may be totally blind and see absolutely nothing, or be legally blind, with serious visual deficiency preventing him from performing the normal activities of daily living. Today, scientists differ on the definition of legal blindness. According to the World Health Organization (WHO), there are sixty-five varying definitions of this state (World Health Organization 1966).

In Jewish law the term "blind person" usually refers to someone who is totally blind. However, "blindness" is sometimes applied to those who lack in one eye, or who suffer from the dimming of vision.

The rabbinical sages use several metaphors to portray a blind person, including "*meor einayim*" meaning "light of the eye" (Hagigah 5b) as a way of easing and softening the harsh situation of the blind, since they recognized the great anguish experienced by the blind. They said there is no greater suffering and no greater pain than blindness (Midrash Shocher Tov #146).

The attitude of Judaism toward the blind since time immemorial has been one of mercy, kindness and help The following are examples of the Jewish attitude to the blind: The Torah commands: "Do not place a stumbling block before the blind" (Leviticus 19:14). The Torah says: "Cursed is he who makes the blind go astray on the way" (Deuteronomy 27:18). When describing appropriate behavior, Job said "I was eyes to the blind" (Job 29:15). A blind person is considered to be of normal intelligence and a normal adult person in all legal matters (Maimonides Zechia u'matana 8:1).

There are dozens of laws in the Jewish codex relevant to the situation of the blind. The following is just a short list of examples to demonstrate the spirit of the Jewish approach, ethically combining divine, holy commandments, the special physical, mental and psychological needs of the blind individual, while also taking into account the needs of the community.

1. A fundamental question about blindness is whether or not it is categorized as a life-threatening condition. This question is relevant for the decision as to whether or not blindness may be treated on the Sabbath. It is widely accepted in *Halachah* that eye diseases are regarded as life endangering conditions, and therefore Sabbath laws are suspended in order to treat these diseases (Maimonides Shabat 2:4).
2. A well known twentieth century rabbinic authority, Rabbi Moshe Feinstein, ruled that the blind were allowed to bring their guide-dog into the synagogue, despite the sanctity of the building into which no other animal was permitted to enter.
3. A *Cohen*, a Jewish priest, is an archetype of wholeness and holiness. Any bodily imperfection prevented him from serving in the Holy Temple. A blind priest—even if only in one eye—could not offer the priestly benediction in order to protect the community from focusing on the disability rather on the benediction, and to protect the priest from inappropriate attention. Nevertheless, in post-Temple times, the *Halachah* makes the following ruling: once the community is familiar

with the individual and the disability has blended into the entity of his personality, he can then fulfill his duties as a priest (Megilah 21a).
4. Especially touching is the following *Halachah*: It is permissible to kindle a light on the Sabbath for a blind woman in labor. Although she derives no direct benefit from the light, it puts her mind at ease when she knows that others can see and will help her. Even if neither she, nor the midwife asks for it, one kindles the light (Maimonides Shabat 2:4).

In this paper we have briefly explored how ethical issues relating to vulnerable populations are approached and resolved within the framework of Jewish bioethics. The paper illustrates the range of sources and methodologies used to determine the appropriate hierarchy of values for various ethical scenarios. Jewish bioethics is based on unchanging sacred principles that can nonetheless be applied to changing contexts through human moral reflection.

King David sings: "Mercy is the cornerstone of the world" (ייבנה חסד עולם, Psalms 89:3). This is a basic, essential principle in Judaism from which this paper has examined vulnerability and disability.

Bibliography

Avot 4:20; 5:24.
Bava Batra 87a; 142b.
Chagigah 14a.
Deuteronomy27:18; 28:50.
Exodus Rabbah 5:12.
Gittin 9a; 23a; 70b.
Hagigah 2b; 4a; 5b.
Isaiah 3:5; 65:20.
Kaplan, H. J., B. J. Sadock, and J. Grab. 1994. *Synopsis of psychiatry, behavioral sciences. Clinical psychiatry*. 7th ed. Baltimore.
Ketubbot 20a.
Kiddushin 32a.
Job 29:15.
Leviticus 19:14, 32.
Leviticus Rabbah 11:8.
Maimonides Hilchot Hamets U'matsa 6:3.
Maimonides, Mishneh Torah, Ishut 13:14; 22:3.
Maimonides Shabat 2:4 Maimonides Zechia u'matana 8:1.
Megilah 21a.
Midrash Shocher Tov #34; #146.
Proverbs 3:16.
Psalms 8:5, 89:3, 92:15.
Reishit Chochma, Shaare Hayirah 15:10.
Responsa Igrot Moshe, Choshen Mishpat, Part 2, 75:7.
Responsa Seridei Eish 3:21.
Responsa Tsits Eliezer 8 15:12, No. 5.
Rosh ha-Shana 29a.
Shevuot 30b.

Sotah 4ab; Sanhedrin 97a.
Strous, R. 2001. Halachic sensitivity of the psychotic individual—The Shoteh, Assia—*Jewish Medical Ethics* 4 (1):30–34.
Shoham-Steiner, E. 2008. *Involuntary marginals*. Jerusalem: Zalman Shazar (Center for Jewish History).
Sontag, S. 1978. Illness as metaphor.
Steinberg, A. 2003. *Encyclopedia of Jewish medical ethics,* 354–376. New York: Feldheim.
Ta'anit 20.
Tractate Berakhot 7a.
Tur, Shulchan Aruch, Yoreh Deah 255:21.
Tur, Orah Hayyim 460.
World Health Organization. 1966. Epidemiology. Vital statistics report 19; 473.
Yerusahlmi Moed Katan 3:1.
Yevamot 112b.
Wikipedia entry: Elderly.
Zechariah 8:4.

Chapter 18
Jewish Perspective on Vulnerable Groups: Women and Children

David Heyd

18.1 General Theoretical Observations

From a metaphysical point of view, every entity or being in the world is vulnerable except for God. Contingency means susceptibility to being changed. Obviously, only changes which in some sense are contrary to the entity's nature, ends or interests are indications of vulnerability.

Thus, rocks are not vulnerable, although they are subject to geologically or climatically induced changes. But plants are vulnerable to pests, and animals—to being prey to other animals. Human beings are vulnerable to constant natural and human hazards. Vulnerability is, therefore, the susceptibility to threats on an entity's very existence and integrity (which may be defined in moral or teleological natural terms; Schroeder and Gefenas (2009)). The etymology of the word is "wound", which is literally an injury to the integrity of the skin, the skin being the protective surface of the body but a very thin one.

If in the wide sense of the term, *all* human beings are vulnerable by being mortal and exposed to ongoing potential attacks or threats by human and natural agents, then we must narrow down the concept of vulnerability so as to cover a *specific* group of human beings who are particularly exposed. There are four principal groups of vulnerable persons in the bioethical sphere which are discussed in detail in this volume (although there are of course many more), which may be matched in two groups: the elderly and the handicapped, on the one hand, and women and children, on the other. The elderly of course tend to suffer more than the general population from physical and mental disabilities, but the concern of this article is with the matching of women and children. The general claim here will be that beyond being two independently weak groups calling for special protection, women and children are in a way mutually vulnerable or exposing each other to particular weaknesses due to their interdependence.

D. Heyd (✉)
The Hebrew University of Jerusalem, Jerusalem, Israel
e-mail: david.heyd@mail.huji.ac.il

J. Tham et al. (eds.), *Religious Perspectives on Human Vulnerability in Bioethics*, Advancing Global Bioethics 2, DOI 10.1007/978-94-017-8736-9_18,

Children are unique in the kind of vulnerability that characterizes them. Unlike the handicapped, children's vulnerability is not a matter of pathological circumstances or bad luck; unlike minorities, children's weakness is not a matter of social discrimination or prejudice. Their vulnerability is natural (that is to say, inevitable), on the one hand, and temporal, on the other. Although every child as such is vulnerable, every child also ends up growing into an independent adult. But the development of human beings into full adults happens to be the longest in the animal world, and hence requires a particularly extended process of care and socialization.

But there is a deeper level in children's vulnerability: their very existence is conditioned by the parents deciding to have them and then to support their biological development in utero. Here vulnerability is mostly social or ethical rather than natural. Since human beings have acquired modern effective means of birth control, including contraception and abortion, children in the wide sense to be immediately specified have become more radically dependent on their parents' will. This explains both why children's vulnerability is unique and why it is deeply connected to the position of women in society and their particular vulnerability.

We may thus describe the concept of children under three categories which are all relevant in the bioethical discussion of their vulnerability:

1. Children before conception
2. Children after conception but before birth
3. Children after birth (till adulthood)

The rest of the first part of the paper will examine each of these categories in some detail. The focus will be on children and the vulnerability of women will be discussed only in the context of their standing as mothers to children (there is obviously a huge literature, both general and Jewish, on the particular susceptibility of women to oppression and discrimination, but I will not deal with it here). The second part of the paper will illustrate the way Jewish law treats children under those three categories in the same order. The third part will shortly touch upon the lesson we can learn from the study of the particularity of the Jewish view about the paradoxical role of human rights in the global discourse of bioethical problems.

The first category is philosophically problematic. It relates to *possible* children rather than to actual young human beings. It refers to a group of unidentifiable individuals who could (hypothetically) exist. There is an intense philosophical debate about the way to analyse the status of such beings. Obviously, un-conceived children have no moral standing and do not enjoy rights or need protection. Calling them "vulnerable" is paradoxical.

Only by a stretch of ontological or theological imagination can we say that since they cannot do anything to bring themselves into existence, they are totally dependent on us and, in that respect, are vulnerable to our decision not to conceive them. However, things become more complex and debatable when children who we know *in advance* will have a genetic defect are conceived. Can *they* claim, after being born, that they were wronged by being created? Here we get entangled in the so-called "wrongful life claims", with which I have dealt extensively elsewhere (Heyd 2009). In this context the question becomes one of bioethical nature, since these law suits typically arise in cases where the child is born with a serious handicap which could have been identified before conception.

The argument is that in such cases the doctors should have advised the parents not to have the child to begin with, and that the parents should have accepted this advice. If they did not—they can be sued by the child born with the defect.

Still, some philosophers, myself included, believe that such claims have no logical foundation and that possible children cannot be "protected from being born." The argument is intricate but its main point is that at the moment the decision regarding procreation is made (even after genetic counselling), the child does not exist and has no identity, let alone interests or rights. Furthermore, it is impossible to form a clear sense (and criteria) for saying that it would be better for a person not to be born at all, despite such psychologically natural feelings as those expressed by the biblical Job and by Sophocles.[1] Even if one can say that dying would be better for an individual than continuing to live, it would make no sense to say that not being born in the first place would have been better for him.

These questions are crucial to the way we address—both legally and morally—the questions of choices regarding the conception of a child whom we know in advance is likely to suffer from a serious disease or handicap Should we avoid bringing such a child into the world, or should we go ahead anyway and plan to take the best care of the child once it is born? These are extremely contentious issues, but, if I am right, they cannot be decided in terms of the rights or the best interests of *the child,* that is to say, they cannot take the principle of the vulnerability of children to guide the decision. Vulnerability does not extend to *existence* (or non-existence). It applies only to individual human beings *once* they are created.[2]

But this conclusion brings to the fore the role of the parents, and particularly of women, on whom the full responsibility for making procreative choices is laid. And on that matter, one can say that women are potential subjects of pressure by husbands, extended family, doctors and society at large. Pressures might work both ways: in more traditional societies they may encourage producing children "at all cost," even when the burden of raising a handicapped child is great and falls primarily on the mother. On the other hand, in liberal societies they might try to push women to avoid having children that fall short of a certain standard of health and "normalcy." In both cases women are put in a sensitive position and their autonomy is curtailed. Unlike possible or potential children, women making choices about such children are vulnerable.

A very typical case of choices about children who have not been conceived yet is the technologically easy method available nowadays of sex selection. Again, it is hard to show how such a choice affects the welfare of the future child and hence it

[1] Oedipus at Colonus, 1224–1228. "Not to be born, by all acclaim, were best."

[2] There are many philosophers today who regard future generations as vulnerable since we can easily harm them but they cannot harm us. This applies to issues of distributive justice and to our duties to preserve enough resources for posterity. It is a complex issue, since all depends whether future people are regarded as just possible (or potential) as in our discussion here, or as given actual individuals who simply do not exist yet but will exist anyway. See my *Genethics* quoted above.

would be strange that such a decision could be made in the light of "the best interests of the child." Potential children, who are not yet conceived, cannot be described as vulnerable to the whims of their parents deciding their gender identity. But if the choice of the sex of the future child (born through IVF) is available to the parents, it is *their* interests which count. Yet in this particular sphere, women often stand in a much weaker position than men.

In many societies the male partner is the side which puts pressure on the woman to have a male progeny and women internalize religious or cultural views which assign sons a higher value than daughters. Laws allowing for sex selection should be very cautious and protect women from male domination in the process of making such a choice, especially since it involves the painful and somewhat risky procedure of IVF, which applies exclusively to the woman.

While the debate about the moral standing of children before conception is mostly very recent, due to technological advances in genetic diagnosis and fertility treatment, the controversy about the second category, that of children between conception and birth, has a long history in theology, law and ethics. The fundamental philosophical difference between preconceived children and fetuses is that while the former lack identity, the latter are clearly identifiable. Or, in traditional logical terms, a preconceived child can be referred to only *de dicto* (e.g. the next child of Mr. and Mrs. X who will be their second), but fetuses can be referred to *de re* (*this* particular future child growing now in his mother's womb). However, we should take note that it is the kind of identity of fetuses and children which is contested. Surely, embryos (at least after the stage of "twinning", around 14 days after conception) have a genetic and physical identity as a continuous bodily matter with a particular organic structure.

And in that sense they are vulnerable to external impact like any other organic creatures. The question is whether they have at that stage *human* identity or *personal* identity, since only on that basis can they be defined as vulnerable in the relevant bioethical sense in which we are interested.

Here our discussion turns from the logical to the metaphysical. There is a wide array of views about what makes human beings what they are, what their essence is: being in God's image, rationality, consciousness, autonomy, speech, etc. The problem is that whatever metaphysical stance we take on this question, human beings do not become what they are from one moment to the other. There are diverse views about the time in which the relevant property which makes us essentially human appears in the process of the development of a child: from the moment of conception, in which a spiritual element in the early embryo is invested through divine intervention; through the creation of the nervous system a few weeks later, which may make the fetus susceptible to pain; to the so-called moment of vitality in which the embryo starts moving by itself; up to the ultimate moment of birth in which the child becomes biologically separated from the mother. We cannot get into that very complex debate about the beginning of the life of a person, but whatever the answer is, it would affect the question of vulnerability. Only when human tissue or a group of cells becomes in some sense human, can it be considered as vulnerable and worthy of particular protection.[3]

The long process of *gestation* leads to the creation of the human person, and this gestation takes place mostly during pregnancy, that is to say in the body of the mother.[4] Here again, the interdependence between women and children is manifest, making both vulnerable not only to the outside world but to each other. The fetus is absolutely dependent for its life and healthy development on the mother; the mother is dependent on the child at least in the sense that her health may be threatened by its existence and, in extreme cases, will be able to live only if the pregnancy is stopped and the child discarded. It is a kind of symbiosis in the literal sense of the word. Fetuses are particularly vulnerable due to their total dependence on their mother's good will; but pregnant mothers are also vulnerable in being held responsible for the development of their children even in cases where their pregnancy is undesired, painful or risky to their life and health.

Finally, the third category—that of children after birth—consists of the most natural cases in the discussion of the vulnerability of children. Here we refer to actual individuals who are biologically independent but in need of close care and protection due to their social, economic, cognitive and psychological dependence on their parents. Children are typically and universally vulnerable to abuse at home, exploitation in the work place, and neglect by a society which in the short term cannot benefit from them. Unlike the previous two categories, children from birth to adulthood are weak and susceptible to harm in ways which are not typically bioethical. However, there are some important bioethical issues regarding children, like their incapacity to give informed consent to medical treatment or to the wrongness of including them in medical experiments, donating organs (e.g. to a sibling), or euthanasia (not having the capacity to give advanced directives in the case of voluntary euthanasia, or not having a profile of presumed preferences or interests which could guide decisions of involuntary euthanasia).

These are the conspicuous bioethical contexts in which children should be regarded as particularly vulnerable and steps taken to protect them by law and regulation.

---

[3] Ronald Dworkin has suggested that the way the abortion debate is conducted should change; and, instead of searching for the moment in pregnancy in which the fetus becomes a human being with full moral standing and protection, we should treat the creation of human beings as a process and the protection of fetuses as a matter of increasing degree. This implies that the concept of vulnerability can itself be applied only gradually, and (despite the possible air of paradox) we can say that the less advanced the fetus is, the less vulnerable! See Dworkin (1993).

[4] There are today periods in which gestation takes place outside the body of the mother, both in the few days of IVF that precede implantation, and in incubators hosting premature babies for a much longer period than ever in the past. These two environments of the development of the human embryo/fetus will definitely extend in time in the not too distant future, which will require rethinking and maybe redrawing the lines between the three categories we are using here (merely planned children, fetuses and children).

But again, women's potential vulnerability is accentuated by children, that is to say, mothers are even more vulnerable than other women due to their special responsibility for raising the children and the particular dependence of children on their mothers. This indeed is mostly a social, rather than biological, condition and hence can in principle be changed on the political and social level. For example more equal parenting of mothers and fathers can empower women, as can the amelioration of work conditions for mothers. But these subjects lie beyond the scope of the bioethical sphere.

18.2 The Particular Jewish Perspective

So far the discussion revolved around theoretical aspects of the subject. As for the specifically Jewish approaches to women's and children's vulnerability in the bioethical field, my comments will inevitably be much less systematic for the simple reason that in the extremely rich Jewish literature on these topics there are deep divisions and unbridgeable disagreements. One should also point out to the novelty of many of the dilemmas regarding assisted parenthood, genetic counselling, prenatal testing, sex selection, contraceptives and many forms of abortion. So although there is a lively debate among Jewish scholars and rabbis in the last few decades about these issues, they have only scarce precedents in the long tradition of Jewish law (*Halacha*).

In this section I will mention some of the Jewish approaches to the issues raised by the three categories of "children" analysed in the first part, one by one.

There is some apocryphal reference to the idea of possible people who have never been born. One midrash (homiletic) about the story of Cain and Abel interprets the verse "your brother's blood cries out to Me from the ground" (Genesis 4:10)[5] as referring to the lives of *all* the possible descendants of Abel (which means half of humanity at that stage). It is as if those poor souls were murdered together with their forefather and are complaining to God for not having received the chance to be born. Or take the story of Amram, Moses' father, who being told of Pharaoh's instruction to kill all Israelite male progeny considered stopping procreating altogether. One of his reasons to continue producing children nevertheless was that by avoiding procreation he would "sentence to death" also all (future) female descendants.[6]

Again, it sounds as if future daughters of Israel are waiting in line to be born and should be given free passage to this world. But as I said, such stories must be understood as metaphorical since the Jewish view is that human beings have no pre-birth existence of any kind. Hence, I believe, that the Jewish view, these stories notwithstanding, does not advance the idea of the right to be born.

[5] All the English translations of the Hebrew Scriptures are taken from Tanakh: The Holy Scriptures (1985).

[6] Babylonian Talmud, tractate "sotah" 12a.

A very famous existential or metaphysical debate takes place between the House of Shamai and the House of Hillel: is it better for man to be born than not to be born? After having argued about it for 2.5 years, they concluded that it would be better for him not to be born, but once born—let him search his soul (examine his deeds).[7] This is a nice and typically Jewish story which starts with a metaphysical (or logical) conundrum but ends in an ironical denial of its importance, directing our attention to the concrete moral goal of doing the best in and with our actual lives.

But when it comes to parental decision about having children, the Jewish tradition is unambiguous. The principle of "be fertile and increase" is not only the very first commandment in the Bible (Genesis 1: 28) but arguably the supreme value of human life and existence—the meaning of it all. The subject of the commandment is controversial (does it include also women or just men) and so is the scope (should one be satisfied with having one boy and one girl or is one expected to have as many children as possible)? But childlessness is in any case considered a sin or a tragedy. This implies that women are vulnerable to the curse of sterility. Remember that the three Matriarchs—Sarah, Rebecca and Rachel—were originally sterile and all in deep despair till they were able to conceive. Furthermore, Jewish law makes women vulnerable in that it allows the husband to divorce his wife unilaterally after 10 years of childless marriage (although today many rabbinical opinions oppose this rule).

The social standing of a woman is dependent on motherhood. Although the commandment to procreate applies strictly only to men, childlessness is a curse and a stigma only on women.

Modern technology of assisted parenthood benefits women with fertility problems and often saves marriages. But it also has the potential of creating a pressure on women to undergo fertility treatment even when it is painful, or stands a very low chance of success. Most rabbis are happy to endorse IVF procedures, surrogate motherhood and, in extreme cases, even sex selection because of the high premium put on procreation. This deep Jewish commitment to the creation of human life definitely stands behind the fact that Israel has the largest number of IVF clinics per capita as well as the first law in the world regulating surrogacy agreements (1996). But typically, and lying with the argument of this article, the interdependence of children and women make the two groups doubly vulnerable.

The modern techniques of overcoming childlessness is both empowering women, adding to their autonomous procreative choices and control, but at the same time also exposing them to pressures from husbands, extended family and social expectations. This is why feminists are divided on the question of the value of fertility technologies.

Moving to the second category, from conception to birth, there is a fairly extensive Jewish literature and ruling on abortion. Originally, the subject was treated (already in the Bible) in narrow legal terms. The only reference to abortion in the Torah tells of a man who in the course of a fight with someone else hits a pregnant woman who consequently miscarriages. The question asked is whom should he

[7] Tractate "eruvin", 13b.

compensate. And the answer is that it is the husband/father. This means that the value of the life of the fetus lies in the satisfaction or benefit it potentially has for the father. Only if the pregnant woman dies as a result of the act of violence, the attacker is accused of murder and is punished by death (Exodus 21: 22–25). The straightforward implication is that causing the death of a fetus is not considered murder and accordingly that the fetus is not regarded as a person in the full sense.

But some scholars, like Maimonides, hold that even if the fetus is accorded full human status and its life is of an equal value to that of the mother's, abortion is justified in cases where the very existence of the fetus threatens the life of the mother. The fetus in this case is considered as *rodef,* that is to say "pursuer" who may be killed as in any act of self defense, even if the pursuer has no ill intention.[8] This way of framing the dilemma of the mother's life against the child's is reminiscent of the famous article by the philosopher Judith Thomson who compared abortion to the mother ridding herself from an unwelcome and dangerous intruder (Thomson 1971).

And indeed, in the later development of Jewish law on abortion the general consensus is that fetuses are different from children after birth and that only after coming out of the mother's uterus does the individual achieve full legal standing with all the protections involved. Thus, in the most clear-cut case of the threat of a continuing pregnancy to the mother's life, there is total agreement throughout the ages that the life of the mother gets absolute priority. In the language of the Mishna,[9] one may actively kill the fetus by amputating its limbs to save the life of the mother (usually in the process of giving birth, but *ipso facto* also in earlier stages of gestation). But when the life of the fetus stands against some other value which is lesser than the life of the mother, there is ongoing disagreement among Jewish authorities. Some say that one should not violate the Sabbath in order to save a fetus; others contest that.

Many allow abortion when the fetus is diagnosed with serious illnesses like Tay Sachs; others would permit an early abortion for more minor illnesses or even in cases of deep depression or mental disturbance of the mother. Regarding very early stages of pregnancy, there is fairly wide agreement that discarding fertilized ova which were found in PGD to suffer from serious genetic defects is permitted, as is the practice of stem cell research and also the procedure of embryo selection in cases of multiple embryo pregnancies. But there are definitely some opposing opinions. The metaphysical assumption behind the more permissive attitude to early abortions is that till the fortieth day the embryo is "mere water", and even after that, it is an integral part of its mother's body rather than an independent individual.

It is noteworthy that the standard reasoning behind the prohibition of abortion does not refer to the moral standing of the fetus and the ascription of personhood to it. Rather it relates to reasons like sperm abuse, the injury to the woman's body and health, the prevention of the future observance of the Sabbath by the person who is now aborted, and the diminution of the image of God (which runs contrary to the

[8] Maimonides. Mishne Torah, Laws of Murder, 1:9.

[9] Mishna. "Ohalot", chapter 7, section 6.

commandment to be fruitful and multiply). Again, this anti-abortion reasoning is different from both the Catholic and contemporary moral arguments which focus on the sanctity of the life of the fetus[10]. In terms of our discussion here, the fetus is not primarily regarded as vulnerable, a weak human being who deserves special care for its own sake. It is protected only to the extent that it is a human being in-the-making.

This might be the reason why Jewish rabbis have little to say about the so-called "wrongful life" cases, in which children sue their negligent doctors for having brought to their creation with a genetic handicap, even in cases when without the negligent act the child would not have been born at all. As mentioned above, human beings have no right not to be born (or for that matter to be born).

It must be emphatically noted that despite all that was said in the previous paragraphs, the Jewish view on abortion is by no means "pro-choice" or "liberal" in the modern sense of the term. The value of human life in general, and the concrete potential of the particular fetus to become a full human being give special protections to embryonic life. Abortion is never taken lightly and is restricted to serious cases of threats to the mother or the child itself. The whole idea of the autonomy of the mother to decide about her body and her life plans is alien to the traditional Jewish thinking about abortion. The whole discussion revolves around the weighting of the real interests (usually health-related) of the mother as against the value of the emerging life of the child. Under the pressure of the religious sectors in Israel, the earlier, more liberal abortion law was changed in 1977, omitting the "social condition" that legitimized abortions for family reasons (like the number of existing children).

Finally, when we consider the third category, that of children after birth till reaching adulthood, the Jewish sources have less to say in the particular fields of bioethical interests in which we are engaged. Informed consent in getting medical treatment, or in taking part in medical experiments, is not a central problem in a religion that does not take seriously the principle of autonomy—in adults, let alone in children. Yet, children are protected in the strongest terms against abuse and physical injury since their lives are equally valuable as those of adults. But their legal and religious standing is kept clearly distinct. There is a large set of rules about the kinds of exemptions children have from fulfilling the commandments and their limited responsibility.

These exemptions are on the one hand a sort of benefit enjoyed by the particular group, but they relegate the group to a weaker social position on the other. All the children's interests are served in the light of the paternalistic judgment of their parents or custodians, which includes all medical decisions. Under the influence of modern liberal views, this is gradually changing, at least in civil Israeli society where children's consent must be obtained in such contexts as medical experiments and with older children even in decisions about medical treatment. Girls under the age of 18 are given the permission to get prescriptions for contraceptives even without their parents' knowledge.

---

[10] For a detailed discussion, see (Steinberg 1998, pp. 74–80).

Adolescents have a say about the medical treatment they get. But these changes have only affected religious rulings very marginally.

As for women (as mothers), they are expected to be homebound, take care of the children and the home in general. There is a traditional saying that the "princess' dignity lies inside", that is to say, the real value of women's life is in the private sphere, at home, primarily raising up the children. Obviously this attitude makes women more dependent on their husbands on all fronts of public life.

To sum up: of all the vulnerable groups that are mentioned in Jewish sources (such as the poor, the foreign resident, the handicapped and the elderly) there is particular prominence of women and children. And since women and children are naturally and, as a matter of strict religious duty, protected by the family, i.e., husbands and fathers, the most vulnerable sub-group lacking protection consists of widows and orphans. They are the by far the weakest in society. The divine passion with which they are protected is captured in the famous verses, in which yet again, women and children are coupled (Exodus 22: 21–23):

> You shall not ill-treat any widow or orphan. If you do mistreat them, I will heed their outcry as soon as they cry out to Me, and My anger shall blaze forth and I will put you to the sword, and your own wives shall become widows and your children orphans.

18.3 The Tension Between Universal Human Rights and Cultural Autonomy

This short exposition of the basic principles underlying the Jewish approach to the ethics of treating women and children as a "combined vulnerable" group raises a general problem facing attempts to reach some global agreement on bioethical issues. It highlights—as the principles of other major religions do—the theoretical issue of culture dependence. Any comparative examination of the way different societies protect vulnerable groups reveals in a compelling way the dependence on cultural traditions and religious faith. Thus, for example, the metaphysics of the human person varies considerably between the Christian conception, the Jewish view and the belief in some Eastern religions: is the person created at the moment of conception? Is there some form of life before present life? Does normative (moral and legal) protection depend on the biological process of the development of the fetus?

Furthermore, the status of women in society varies from one tradition to the other, especially when it comes to reproduction (as we have partly seen in the previous section). This heterogeneity gives rise to the problem not only of cross-cultural mutual understanding but also of practical coordination in the global sphere of bioethical norms and principles for protecting vulnerable groups. The post-WWII period strives to achieve consensus on these matters in a historically unprecedented way. The main vehicle for a global agreement has been the United Nations and its various conventions.

But this move by the international community to create some global common ground has proven to involve a fundamental *paradox*. The only framework which has been suggested already in the beginning of the process, in documents such as the UN Declaration on Human Rights, inescapably appeals to the notion of *human rights*. Nowadays there are more and more voices which express a critical attitude towards the adoption of the discourse of rights as the dominant language in which bioethical practice is judged, since they claim that this is a "Western" tradition of liberal individualism typical of the last four centuries in the European tradition. This critique is indeed historically valid, but misses the point. The reason why the discourse of human rights is the best candidate for a common denominator in the international discussion of justice is that it is the only discourse which is inherently *universal*.

It approaches the normative problems as pertaining to human beings *qua* human beings (rather than members of a particular religious or political community). Liberalism offers (or at least tries to offer) a morally neutral framework for some basic principles, and hence hopes to avoid the culture-dependent dimension of particular societies. Human rights have proven to be a very powerful tool in the hands of individuals and sub-groups who have a complaint against their governments or rulers in domestic societies, but it is equally used by whole nations on the international level, particularly by societies which have been victim to colonial oppression. Self-determination, economic justice and political participation are all examples of the appeal to human and cultural rights even by societies in whose culture rights have not played a significant role.

It is often claimed by historians of ideas that the concept of rights emerged within the so-called Judeo-Christian tradition. This is to some extent factually true. But as I tried to show in the previous section, the Jewish tradition—past and present—is not typically geared to human rights and their underlying conception of the person. It is more "communitarian" than individualist and more paternalistic than autonomy-based. In the case of reproductive choices, the relevant considerations are neither the rights of the future child nor the free choice of the woman and her control over her body.

And although the protection of embryos is often interpreted today in terms of the embryo's right to life, the original Christian doctrine speaks of the abstract value of the sanctity of life rather than the protected interests of the individual embryo in coming to life. And one may generalize also to other vulnerable groups in society, such as the elderly and the handicapped, by arguing that it is charity, respect, recognition of being created in God's image that justify the protection of these groups rather than their inalienable or natural rights.[11]

There is thus an inherent tension between the plea for recognizing the particular cultural attitude to bioethical dilemmas and the widespread appeal to do so by using the language of rights, which is often alien to the fundamental views of that culture. One interesting attempt to reconcile the liberal language of rights with the communitarian language of traditional religious norms is to shift focus to the autonomy

[11] It might look quite striking that in the seminal (and pioneering) work of Immanuel Jakobovits (1959), the term "rights" does not appear in the index.

of communities rather than individuals[12] (Kymlicka 1989). Thus, even cultures that do not hold their members as having individual rights as supreme normative constraints deserve respect by other cultures; that is to say, they have the right to be left alone in managing their ethical affairs. Accordingly, liberal societies, or indeed the international community as a whole, owe cultural groups and religious communities a large measure of freedom in the way they treat their own vulnerable groups according to their traditional norms. However, this second-order liberal principle of respecting communal (rather than individual) rights is vulnerable to the question of the right way to deal with cases in which there is a contradiction between the patently illiberal way the community treats, e.g. its female members and their universal rights, as human beings.

Unfortunately, these contradictions are widespread and create the well-known problems regarding the right response to practices such as female circumcision (which happens to illustrate so painfully well the interdependence of the vulnerability of children and women!). Another typical example for this contradiction is the so-called "right of exit", according to which cultural communities should get autonomy in organizing their social norms and lives as long as every individual is free at any time to leave the community. This sounds like a plausible principle, but it can hardly apply to the vulnerable groups of women and children, the elderly and the sick, who by their very nature *lack* that freedom.

But as in matters of political and economic justice in the global sphere, it seems that there is no alternative in international cooperation than the language of rights—of children, women, the handicapped and the elderly—as is the case with religious toleration, political independence, economic equality and future generations. The typically UN rhetoric of rights, despite being suspected as itself "parochial" (namely, "just Western") provides a common universal roof under which particular perspectives may find their expression and ground their plea for recognition—whether on the individual level or the communal.

References

Dworkin, R. 1993. *Life's dominion*. New York: Knopf.

Heyd, D. 1992. *Genethics: Moral issues in the creation of people*. Berkeley: University of California Press.

Jakobovits, I. 1959. *Jewish medical ethics*. New York: Bloch.

Kymlicka, W. 1989. *Liberalism, community and culture*. Oxford: Clarendon.

Schroeder, D., and E. Gefenas. 2009. The intractability of the nonidentity problem. In *Harming future persons,* eds. M. Roberts and D. Wasserman, 3–25. Dordrecht: Springer.

Steinberg, A. 1998. Halakhic and medical encyclopedia, vol. 2. Jerusalem: Schlesinger Institute.

1985. Tanakh: *The holy scriptures*. Philadelphia: Jewish Publication Society.

Thomson, J. J. 1971. A defense of abortion. *Philosophy and Public Affairs* 1:47–66.

[12] The first and best known philosopher to have articulated this idea is Will Kymlicka (1989) in his series of publications starting from Liberalism, Community and Culture.

Chapter 19
Lessons Learned

Joseph Tham

It has been quite a stimulating experience to read the many different papers, with such a variety of positions and backgrounds, on the rather novel topic of vulnerability in bioethics. We have been entrusted with the task to make a summary of the different papers, to mention some of the lessons learnt as well as the challenges that still lie before us.

First, we will briefly summarize the different positions of the six religious groups based on the papers of this volume stressing the areas of convergence and the differences. Second, we will discuss the various questions raised in these papers and the lessons learned from them. Here we will proceed to look at the contentious concept of vulnerability in bioethics and the challenges regarding its status, content and scope. Then, we will look at the religious dimension and how it offers an added value to the discussion on vulnerability. In particular, all religious traditions seem to point to a more realistic assessment of the vulnerable human condition.

This is often contrasted with an overly individualistic understanding of autonomy. Responsibility toward the other is discussed, as well as the prophetic dimension of religions in these issues. Finally, we will look at the ever-present tension between universal human rights and local (religious) traditions.

19.1 Different Religious Perspectives

The papers presented study the question of vulnerability in bioethics and how religious traditions view this topic in the same way. Some authors analyze the question of vulnerability more from a philosophical perspective and only then examine them from a religious angle. Others look at the problem of vulnerability from a more specific question like informed consent, while others opt to paint in very wide brushstrokes the general attitudes of care from their religious traditions. The following summaries will be grouped more along the line of contiguous traditions.

J. Tham (✉)
School of Bioethics, Ateneo Pontificio Regina Apostolorum, Rome, Italy
e-mail: jtham@legionaries.org

J. Tham et al. (eds.), *Religious Perspectives on Human Vulnerability in Bioethics*, Advancing Global Bioethics 2, DOI 10.1007/978-94-017-8736-9_19,

Hence, we will first look at the "Western" tradition, meaning the so-called Judeo-Christian approaches. This is followed by Islam due to its monotheistic affinity. Then we will look at the Eastern religions, starting with Hinduism, followed by Buddhism, with their particular emphasis on mysticism. We conclude with a look at Confucianism which is more ethical, and culturally based and less "religious."

Judaism emphasizes the primacy of God and the condition of human beings as creatures who as partners can cooperate with the Creator. The emphasis on human dignity is based on the theological understanding of being created in the image of God. Hence, human vulnerability is a consequence of our condition of being created. Barilan noted that whereas human dignity forms the basis of human rights whereby all of us deserve equal treatment, the vulnerable deserve special protection. Hence, there is special sensitivity towards injustice, discrimination, prejudice, and being ostracized because one is different or belongs to a minority.

At the same time, Heyd notes that Judaism is more communitarian than individualist, and more paternalistic than autonomy-based in its ethical worldview. This tension between traditional communitarianism and liberal individualism dominant in the rights discourse is not easily reconciled. In Judaism, there are also a variety of opinions regarding how to resolve ethical dilemmas. Even though there is a common basis of sacred texts, without a central authoritative body there can be, and indeed are, various interpretations or opinions from different schools or rabbis. Halevy notes that in Judaism a flexibility to adapt traditions to the changing cultural and scientific technical environment exists.

The other papers also indicate openness to philosophical inquiries as an adjunct to religious input. The paper by Barilan is especially insightful about the vulnerability of decision-making process and that, in spite of autonomy, the right answer can sometimes be elusive, which points us back to the finitude of the human condition in the face of the eternal invulnerable God.

The three papers on Christianity came from Orthodox (Tzitzis), Protestant (Ucko) and Catholic (Gallagher) authors, and were in general agreement with the Jewish biblical perspective on the centrality of the human person as bearer of dignity and rights. In addition, Christians heed the command of Christ that in our dealings with strangers, especially the most vulnerable, we are in reality serving Christ. Hence, the effort to be merciful to those who are less fortunate is not just humanitarian service or philanthropy, but is raised to a higher level imbued with a deep religious significance. There is no branding the other a stranger, but all individuals are considered neighbors who demand one's compassion and solidarity. Ucko and Gallagher reiterated the duty of Christians as advocates and prophets on issues of justice and health care delivery.

Thus caring for those who suffer, the weak, the poor and the disenfranchised is part and parcel of the Christian vocation to love, or *agape*, as Tzitzis eloquently described. Yet this high calling to *agapeic* love is also bound to contradictory selfish tendencies. As Blaise Pascal mentions, human existence is a paradox: We are at once called to greatness and yet bogged down by our own wretchedness. Our human vulnerability and brokenness, as Tzitzis mentioned, can be assimilated into Christ's brokenness in the Eucharist. These authors are also discontented with the current trajectory of human rights.

Even though the human rights tradition can certainly be traced back to its Judeo-Christian roots, the so called Western liberal approach may not be totally compatible with the Christian worldview. Tzitzis emphasized that human rights can only find their true meaning when it accepts the ontology of man. Hence, he is against the nihilistic tendency of a pure autonomous self which negates the attribution of life to the most vulnerable, especially at the beginning and end of life (i.e., embryo destruction and euthanasia).

Even though Islam is also a monotheistic faith that traces its roots back to Abraham, its relationship with human rights claims are much more thorny than Judaism's and Christianity's. Atighetchi's paper opens with a provocative description of the assault and invasion of Western laws on the Islamic legal culture. In fact, according to Islam, the universality of human rights is reason-based and so must be subordinate to universality of Islamic revelation which is supernatural. In this perspective, Koranic revelation and their authentic Islamic interpreters cannot be challenged by external critiques, including perceived infringements of human rights. Apparently, equality is understood differently in Islam, as noted in its attitude towards women and non-Islamic religious minorities.

However, according to the Koran, mercy and just treatment should be extended to the weak, poor, and minorities of the society. Atighetchi also relates the diminished decision-making capacity of women that is often conditioned by males and other family members, a practice similar to Hindu and Confucian practices. According to Abu Sway, the Islamic attitude toward disease and suffering is imbued with religious significance. The Muslim believer must acknowledge the supreme wisdom of Allah and conform to his will, while recognizing his rightful place of nothingness and contingency. This explains the common recourse to faith healing and favoring familial-spiritual methods in the treatment of mental illnesses.

Atighetchi illustrates the difficulty in adapting to change of culture and scientific or medical advances with the example of pain management and palliative care. In fact, the challenges of traditional cultures facing the advances of modernity, be they technological in the case of pain management or cultural in the debate on human rights, are also experienced by other religious groups. However, the monolithic nature of Islam has sharpened this contrast more noticeably.

Desai's paper on Hinduism also elucidates the problem faced by India as it confronts modernity. In the past, Hinduism provided networks of relations to care for and offer protection to those who are vulnerable, including women, children, and the handicapped. As Muthuswamy attests about the Hindu woman, "Father protects (her) in childhood, husband protects (her) in youth, and sons protect (her) in old age. A woman cannot be left unprotected." However, there has been a gradual breakdown of these circles of protection and guardianship in modern India due to migration, isolation, and limited access of care.

India also has to face the daunting challenge of human rights equality concerns, since as Desai has noted elsewhere, inequality is ingrained in the teachings of a Hinduism that has produced the caste system and the untouchables. However, unlike Islam, Hinduism is much more fluid and capable of adaptation to different cultural strands. With regards to the vision of disease, the Hindu approach is holistic, allowing for the body and psyche to influence each other. This spiritualized understanding of health encourages psychosomatic treatments and exercise such as yoga practices. At the same time, the Hindu understanding of sin and its recompense with reincarnation allows for a theology of suffering and sickness as a means of purification and perfection.

Similarly, both papers on Buddhism agree on the fact that vulnerability is not something totally negative, but the essence of human life. This comes from the particular understanding that inevitably life is full of suffering. Suffering points to the illusory nature of life, and pushes the practitioner towards the ultimate goal of nirvana where suffering disappears. The Buddhist attitude towards vulnerability is summarized by interdependence and compassion. Interdependence within the person means that one cannot separate one's ideal self from the diseased self. The acceptance of our existence, prone to disease, suffering, and death, is a form of enlightenment that gives peace and comfort.

In addition, everything in nature, humans included, actually form a web of relations and of dependence on each other. Compassion arises from the awareness of this interdependence. When we observe someone else suffering, this suffering is not alien to me, because we are interrelated. This realization leads to sharing, communion, and ultimately compassion. Compassion is a result of our shared common vulnerable condition as humans. To achieve this awareness requires constant meditative practices.

While the religious tenets of Buddhism can be held more as a way of life, this is much more evident in Confucianism, which many consider, strictly speaking, not to be a religion. Lo's contribution takes great issue with the IBC document on vulnerability that places self-determination as an unsurpassable human right in medical decision-making. In contrast, he offers family co-determination according to the Confucian perspective as the best means to protect vulnerable members. The Confucian family-centered vision is an antidote to liberal individualism, which Lo finds inadequate as a means to protect the vulnerable.

Chan supplements Confucian treatment or attitude towards those outside the family with the virtue of *ren*. Correspondingly, *ren* is graduated benevolence or humaneness that begins within the family and extends outward towards larger circles of society. *Ren* is a personal virtue that extends one's responsibility to care for those who are vulnerable. Chan believes that the liberal understanding of justice, as exemplified by Rawls, cannot account for the reason we have duties toward those who are vulnerable, as presupposed by the Confucian concept.

19.2 The Concept of Vulnerability

Even though we have a general intuition about the care of those who are vulnerable, when we try to pin down the concept, we run into difficulties. Henk ten Have, who headed this effort to draft the UNESCO Declaration on Bioethics and Human Rights, explains that vulnerability is a new concept in bioethics. Just as in most international documents that have attempted to define terms such as "person" or "justice," there are bound to be controversies in the definition of "vulnerability."

Ten Have provides a historical perspective of the evolution of vulnerability in the Belmont Report, CIMOS, Declaration of Helsinki, and finally the UNESCO Declaration. The rise of its importance from special consideration to guideline, to now becoming a principle is noteworthy precisely because there is a surplus of bioethical terms like justice, beneficence, dignity, autonomy and the like. Perhaps this increase in prominence is due to globalization, where vulnerability is more agreeable and acceptable than autonomy in non-Western traditions. The fact that the principle of vulnerability was quickly approved in the Declaration without debates, as Miranda recounts, due to the fact that the delegates were tired after the long debates on informed consent and were worried about the need to protect the most vulnerable of society.

However, as ten Have points out, there are still unsettled issues in terms of status, content, scope and application.

Here, contributions from other authors are helpful in bringing the problems to the fore. Tarasco is very critical of the notion of vulnerability in health care ethics. She believes that vulnerability is too elastic a term to be of use, since it is always relative to some standard. In fact, it is just another word for risk. She prefers the word fragility which better describes our common human condition. At the root of this complaint is another debate about the definition of health.

According to the definition of WHO that: "Health is a state of complete physical, mental and social well-being and not merely the absence of disease or infirmity." While this seems innocuous at first sight, there are those who are unhappy with the inclusion of social dimension into a strictly medical dimension of health. In this light, we better understand why Tarasco is concerned about the danger of ideological use of the term vulnerability to arouse sympathies, because such use is not based on true medical conditions. In this light, when health becomes a product of desire, autonomy would mean the right to fulfill those desires, and vulnerability is then based on a subjective lack of fulfillment, instead of the objective needs of the good of a person. At the heart of this critique is a religious concept of the person as having a tendency toward certain objective ontological goods.

This stands out against a utilitarian understanding of the person promoted in a secular and liberal environment.

Regarding its content, vulnerability is historically seen as a lack or a compromised ability of certain individuals or groups to give full consent to medical interventions; or as better stated in the 1993 CIOMS document, "a substantial incapacity to protect one's own interests." This incapacity can be caused by a variety of situations—financial, intellectual, social, educational, physical, etc.—that makes the person thus affected vulnerable. Hence, it appears that the concept of vulnerability is closely tied to the principle of autonomy with its perceived diminishment. Ten Have further notes that it is also tied to the principle of justice: those who are less capable of defending themselves, and therefore more vulnerable, should have the same access to aid, or even more access, than the rest. Yet, in the end, there seems to be a sense that framing bioethics only in terms of autonomy is insufficient.

He calls for the need for positive demands: "Bioethics is more than respecting individual choices and personal autonomy: it aims at safeguarding care for other persons."

Barilan also observes, "Autonomy is at the heart of vulnerability," but later adds, "So is the conceit of realizing our autonomy without acknowledgment of our vulnerabilities." His paper poignantly describes the tragic paradox of a human being, who is vulnerable *in se*, and any attempt to render him less so is also prone to vulnerability and possible defeat. Barilan calls this the "interdependence of vulnerabilities" which is present within the same person (the mind's ability to predict bodily disease), in his relations with those around him (perceptions, expectations, and self-esteem), and with society at large. Inherent in our vulnerable conditions are all the ambiguities contained in the inability to make the right decision, and the bad consequences that thus ensue. Hence, dignity and autonomy meet at the deepest level of the embodied person's self-identity and conscience, which appeal for protection.

Finally, there is the challenge of demarcating the scope of vulnerability, which can be applied not only to individuals but to families and other groups, communities and populations. At first sight, it appears there can be three types of vulnerability.

First, biological vulnerability refers to a state of corporal vulnerability based on innate traits, fragility, and threats to the person. Second, social vulnerability is derived from conditions resulting from war, crimes, prejudice, hospitalization and poverty. Third, cultural vulnerability can be provoked by particular traditions and conceptions of certain cultures that categorize, generally or specifically, these individuals or groups as vulnerable. In our discussions during the workshop, Lo was insistent that we need to distinguish the first type of vulnerability which is an inherent condition, from the second or third types which are caused by external factors.

For instance, children and the handicapped would belong to the first type, whereas women would fall into the latter. Old age would fall somewhere in between since it is both an inherent condition and receives an accompanying societal treatment. In that sense, while the first type of vulnerability requires great protection and care, the latter requires a change in the environment whereby these groups are no longer discriminated against or stigmatized.

Speaking about stigma, there is also the paradox that labeling certain individuals and groups as vulnerable can itself be a cause of furthering their vulnerability. Ucko speaks of the propagation of the negative image of victimhood that can become a dire self-fulfilling prophesy. Others find the very broad inclusion of a group (e.g., women) as vulnerable to be derogatory and discriminatory. It is therefore interesting to note that while earlier 2009 and 2010 drafts of the IBC Report on the Principle of Human Vulnerability provided a list of vulnerable groups, the final report in 2011 has eliminated these categories and instead only offers examples and cases.

A third area of contention regards who should be considered vulnerable.

Christians and, in particular, Catholics would insist on inclusion of embryos as the most vulnerable member of the human family, as noted by Miranda, Tarasco and Tzitzis. Barilan sees the homosexual and transsexual persons as a neglected vulnerable group that has been discriminated against by religious groups. These contrasting visions are a reflection of the on-going cultural debate about what is constitutive of human nature and the human good, which in turn depends on different philosophical and theological starting points.

19.3 Added Dimension of Religious Perspectives on Vulnerability

How religious perspectives have enriched the understanding of vulnerability is our next topic of consideration. As mentioned in the introduction of this volume, this format of engaging different religious traditions to dialogue and face each other is still relatively new. A space has been opened up for this to take place, but, as in any fruitful conversation, time is needed for trust to develop and openness to expand. In this work, we have a snapshot of the different traditions and their sensitivities with regards to these bioethical issues.

An interesting observation is that groups with a longer history of bioethical reflection tend to be more mature in their reflection and are not afraid of self-criticism (e.g., Christianity and Judaism). Other traditions which have a shorter period of engagement in these issues tend to approach the debates with greater caution, and are more critical of external agents (such as liberalism) rather than practices in their own tradition (e.g., Islam, Hinduism, Confucianism). This is not meant to be a disparaging remark, but a normal progression that comes with maturity of the discipline. One complaint of bioethics today is its attention to dilemma-solving. Ironically, many problems cannot have solutions unless the deeper questions are addressed, problems which largely touch on the *telos* of human life, human good, human nature, and medicine.

World religions have long-standing traditions of reflection on these profound matters, and it would be a great failure, if not foolishness, for us to ignore this source of wisdom. The wisdom of religion may not offer solutions to dilemmas, but it can provide the needed framework for open discussion. Religion can therefore offer an "added value" to deepen the bioethical conversation. One very special contribution from the religious perspective on the subject of vulnerability regards our human condition. In some way, the religious vision understands and accepts humans as conditioned, imperfect, frail, and contingent beings who cannot escape the ultimate reality of suffering and death. In dealing with the handicapped Ucko asks whether, from a theological point of view, "disability is a gift from God rather than a limiting condition."

Clearly there is a tendency in the Bible to see imperfection or disease as the result of sin. Nevertheless, is absolute perfection the ideal to strive for? Isn't imperfection a fact of life?

The story of Rabbi Elazar's encounter with an ugly man as recounted by Halevy is striking in this respect. Why did God make something imperfect? Religious wisdom, however, wants to recall that disability, suffering, and sin are all part of our vulnerable nature. Imperfection is built into our human condition, and it would be hubris to ignore this reality, perhaps even an injustice to try too hard to eliminate all blemishes in our midst. For the Buddhist especially, vulnerability is not something evil to be suppressed, but a part of our being human, and hence something that needs to be embraced.

This vision is somewhat jarring to the secular mindset, which champions the ability to manipulate and exploit nature and control destiny through the omnipotence of technology. Modernity finds it difficult to comprehend the traditional religious concepts of suffering, patience, and acceptance of our limitations and mortality. Instead, it employs all the means at its disposal to fight against suffering and death. Sometimes, however, the cost can be too high, for example, when we eliminate those that are less than perfect at the beginning of life, or can no longer contribute effectively to society at the end of life. In contrast, Tarasco reminds us that not all desires are absolute, and we should not desire things that are contrary to reason, truth, and the common good.

Buddhism goes even further, positing human desire as the origin of all suffering and enlightenment in the realization that all desires are illusory. In a way, these visions can provide the needed "alternative imaginations" to principles or models of bioethics, for believers and non-believers alike. These imaginations may even provoke personal transformation and social renewal. Another area in which religious traditions can supplement and enrich the concept of vulnerability comes from a general complaint of an overly individualistic conception of the human person in the UNESCO Declaration. This grievance is more pronounced from the authors from the Orient.

Buddhism warns about the illusion of autonomy, since the self is ultimately void or empty. In this sense, we are all vulnerable to suffering and death, a condition that we have to accept not as a negative fact but as a truth of life. Hongladarom mentions how this vision of radical human vulnerability can act as a counterweight to rights-based individualism. Hinduism also runs into difficulties with the problem of informed consent and autonomy in the treatment of women, children, lower castes, the mentally ill, and the poor, etc. The solution lies in setting up network of relationships based on kinship or another form of organization to offer protection

For the Confucians, the self is seen only in relation to the family. Thus, the family is the first bulwark for the vulnerable. Lo quotes a passage in his text that is illustrative of this opinion: "[If] I have a trustworthy family and a supportive circle of friends, I would prefer to endure the outcome if they 'err' in predicting my preferences, or even if they choose to ignore my preferences other than the preference for family decision-making, rather than to remove from them the opportunity and the burden of making the choices." The importance of family heads or male tutors to represent or make substituted judgments for women or weaker members of society are also echoed in the papers by Islamic (Atighetchi), Jewish (Halevy) and Christian (Tzitzis) authors.

Other authors (Zhang, Abu Sway, Ucko, Tzitzis) directly or indirectly made reference to the "face of the Other" of Jewish philosopher Emmanuel Levinas. In confronting the naked face of the Other, different from me and yet vulnerable like me, an ethical demand of response is elicited. One cannot ignore the plea of the Other who challenges us face-to-face. The Other, a stranger who shares my humanity, exacts from me a certain responsibility to respect his dignity once I am aware of our interconnectedness. This is the core of all religious commitment to the vulnerable, with the different names of *agape* or charity, neighborly love, solidarity, *visheshdharma*, *ren*, *karunā* or compassion, and mercy or *hesed.*

Related to this is the prophetic task of religion, which can remind society of the weightier questions of common good, justice, and care for the poor and the voiceless. Hence, religions are often called to testify to the supremacy of God when false expectations of technology and manipulation of human lives are brushed over. Religion's prophetic task can help in the discovery of oft-ignored problems in bioethics, possibly because of oversight, but more likely because of complicity and vested interests. Once the root causes are disclosed, the prophet must speak out and convey an alternative vision, consisting not so much in articulating concrete changes in policies, but alerting people to neglected wrongs and injustices. The prophetic voice of religion must call into question unjust practices that perpetuate the scandalous situation and call to accountability those in a position to effect change.

A few examples of these prophetic challenges in these papers mention the unfair allocation of health resources in the world, marginalization of migrant populations, the problem of access of medical care in India, problems of coercion and sexual exploitation, the lack of appropriate palliative care, and the possible misuse of regenerative, neuro, and genetic technologies by pharmaceuticals and multinationals, etc.

Religious communities can finally teach modern medicine this truth: to be present to those who are weak by accepting our finitude and vulnerabilities. Vulnerability can be a common ground where both patients and physicians could meet in the healing encounter. It means that physicians, too, should acknowledge their weaknesses and limitedness, despite the armamentarium offered in modern medicine. Finally, religious communities can become the seedbed for growth in virtues, and in character, faith and holiness. For both the patient and the healer, the character development of the moral agent has not been adequately addressed in contemporary bioethics.

19.4 Universal Human Rights vs. Local Cultural Diversity

Last but not least, we have come to the thorny issue of the tension between the claims of universal rights and respect for local (religious) customs. On the one hand, the international post-war agreement to establish certain commonly held human rights has become a powerful instrument to prevent future abuses and conflicts. These human rights are universal because they pertain to human beings as human beings.

On the other hand, a paradox arises when recognition of the rights of particular cultural attitudes clashes with universal rights. This tension is noted in Article 12 of the UNESCO Declaration on Bioethics and Human Rights: "The importance of cultural diversity and pluralism should be given due regard. However, such considerations are not to be invoked to infringe upon human dignity, human rights and fundamental freedoms, nor upon the principles set out in this Declaration, nor to limit their scope." Lo's paper criticizing the IBC conception of informed consent exemplifies this inherent tension between rights language and cultural understandings of protecting the vulnerable in one's own tradition. This opposition is also noted with varying degrees in all other religious traditions, in part but not exclusively, due to the perceived liberal conception of humans that could be at odds with the religious vision. This debate also became evident in the workshop.

Those from the Western tradition defend the universality of human rights enshrined in the UNESCO Declaration because it was signed by all the member states. However, other participants, especially from the Orient, did not find this consensus to constitute a measure of acceptability. Heyd suggested in his paper a possible solution with the "autonomy of communities," but only time will tell if this theory will work. The fact that there is still a lack of agreement on the nature and scope of human rights in relation to local cultures has propelled us to explore this subject matter in our next workshop on "Bioethics, Multiculturalism and Religion." As we saw in the introduction, the call to an intercultural dimension in bioethics requiring a tradition-constituted conversation that MacIntyre calls for is still a long road to be traveled.

But as an old Chinese saying goes, "The journey of ten thousand miles begins with the first step." We believe that we have taken a small step with this workshop and publication. We hope and pray that future steps will be taken by others to join us in this animated and fruitful dialogue.

Chapter 20
Appendices

Joseph Tham, Alberto Garcia and Gonzalo Miranda

20.1 Appendix 1

20.1.1 Universal Declaration on Bioethics and Human Rights[1]

The General Conference

Conscious of the unique capacity of human beings to reflect upon their own existence and on their environment, to perceive injustice, to avoid danger, to assume responsibility, to seek cooperation and to exhibit the moral sense that gives expression to ethical principles.

Reflecting on the rapid developments in science and technology, which increasingly affect our understanding of life and life itself, resulting in a strong demand for a global response to the ethical implications of such developments.

Recognizing that ethical issues raised by the rapid advances in science and their technological applications should be examined with due respect to the dignity of the human person and universal respect for, and observance of, human rights and fundamental freedoms.

[1] http://portal.unesco.org/en/ev.php-URL_ID=31058&URL_DO=DO_TOPIC&URL_SECTION=201.html.

J. Tham (✉) · G. Miranda
School of Bioethics, Ateneo Pontificio Regina Apostolorum, Rome, Italy
e-mail: jtham@legionaries.org

G. Miranda
e-mail: g.miranda@arcol.org

A. Garcia
UNESCO Chair in Bioethics and Human Rights, Rome, Italy
e-mail: agg@arcol.org

J. Tham et al. (eds.), *Religious Perspectives on Human Vulnerability in Bioethics*,
Advancing Global Bioethics 2, DOI 10.1007/978-94-017-8736-9_20,

Resolving that it is necessary and timely for the international community to state universal principles that will provide a foundation for humanity's response to the ever-increasing dilemmas and controversies that science and technology present for humankind and for the environment.

Recalling the Universal Declaration of Human Rights of 10 December 1948, the Universal Declaration on the Human Genome and Human Rights adopted by the General Conference of UNESCO on 11 November 1997 and the International Declaration on Human Genetic Data adopted by the General Conference of UNESCO on 16 October 2003.

Noting the United Nations International Covenant on Economic, Social and Cultural Rights and the International Covenant on Civil and Political Rights of 16 December 1966, the United Nations International Convention on the Elimination of All Forms of Racial Discrimination of 21 December 1965, the United Nations Convention on the Elimination of All Forms of Discrimination against Women of 18 December 1979, the United Nations Convention on the Rights of the Child of 20 November 1989, the United Nations Convention on Biological Diversity of 5 June 1992, the Standard Rules on the Equalization of Opportunities for Persons with Disabilities adopted by the General Assembly of the United Nations in 1993, the UNESCO Recommendation on the Status of Scientific Researchers of 20 November 1974, the UNESCO Declaration on Race and Racial Prejudice of 27 November 1978, the UNESCO Declaration on the Responsibilities of the Present Generations Towards Future Generations of 12 November 1997, the UNESCO Universal Declaration on Cultural Diversity of 2 November 2001, the ILO Convention 169 concerning Indigenous and Tribal Peoples in Independent Countries of 27 June 1989, the International Treaty on Plant Genetic Resources for Food and Agriculture which was adopted by the FAO Conference on 3 November 2001 and entered into force on 29 June 2004, the Agreement on Trade-Related Aspects of Intellectual Property Rights (TRIPS) annexed to the Marrakech Agreement establishing the World Trade Organization, which entered into force on 1 January 1995, the Doha Declaration on the TRIPS Agreement and Public Health of 14 November 2001 and other relevant international instruments adopted by the United Nations and the specialized agencies of the United Nations system, in particular the Food and Agriculture Organization of the United Nations (FAO) and the World Health Organization (WHO).

Also noting international and regional instruments in the field of bioethics, including the Convention for the Protection of Human Rights and Dignity of the Human Being with regard to the Application of Biology and Medicine: Convention on Human Rights and Biomedicine of the Council of Europe, which was adopted in 1997 and entered into force in 1999, together with its Additional Protocols, as well as national legislation and regulations in the field of bioethics and the international and regional codes of conduct and guidelines and other texts in the field of bioethics, such as the Declaration of Helsinki of the World Medical Association on Ethical Principles for Medical Research Involving Human Subjects, adopted in 1964 and amended in 1975, 1983, 1989, 1996 and 2000 and the International Ethical Guidelines for Biomedical Research Involving Human Subjects of the Council for International Organizations of Medical Sciences, adopted in 1982 and amended in 1993 and 2002.

Recognizing that this Declaration is to be understood in a manner consistent with domestic and international law in conformity with human rights law.

Recalling the Constitution of UNESCO adopted on 16 November 1945.

Considering UNESCO's role in identifying universal principles based on shared ethical values to guide scientific and technological development and social transformation in order to identify emerging challenges in science and technology taking into account the responsibility of the present generations towards future generations, and that questions of bioethics, which necessarily have an international dimension, should be treated as a whole, drawing on the principles already stated in the Universal Declaration on the Human Genome and Human Rights and the International Declaration on Human Genetic Data and taking account not only of the current scientific context but also of future developments.

Aware that human beings are an integral part of the biosphere, with an important role in protecting one another and other forms of life, in particular animals.

Recognizing that, based on the freedom of science and research, scientific and technological developments have been, and can be, of great benefit to humankind in increasing, inter alia, life expectancy and improving the quality of life, and emphasizing that such developments should always seek to promote the welfare of individuals, families, groups or communities and humankind as a whole in the recognition of the dignity of the human person and universal respect for, and observance of, human rights and fundamental freedoms.

Recognizing that health does not depend solely on scientific and technological research developments but also on psychosocial and cultural factors.

Also recognizing that decisions regarding ethical issues in medicine, life sciences and associated technologies may have an impact on individuals, families, groups or communities and humankind as a whole.

Bearing in mind that cultural diversity, as a source of exchange, innovation and creativity, is necessary to humankind and, in this sense, is the common heritage of humanity, but emphasizing that it may not be invoked at the expense of human rights and fundamental freedoms.

Also bearing in mind that a person's identity includes biological, psychological, social, cultural and spiritual dimensions.

Recognizing that unethical scientific and technological conduct has had a particular impact on indigenous and local communities.

Convinced that moral sensitivity and ethical reflection should be an integral part of the process of scientific and technological developments and that bioethics should play a predominant role in the choices that need to be made concerning issues arising from such developments.

Considering the desirability of developing new approaches to social responsibility to ensure that progress in science and technology contributes to justice, equity and to the interest of humanity.

Recognizing that an important way to evaluate social realities and achieve equity is to pay attention to the position of women.

Stressing the need to reinforce international cooperation in the field of bioethics, taking into account, in particular, the special needs of developing countries, indigenous communities and vulnerable populations.

Considering that all human beings, without distinction, should benefit from the same high ethical standards in medicine and life science research.

Proclaims the principles that follow and adopts the present Declaration.

20.1.2 General Provisions

Article 1—Scope

1. This Declaration addresses ethical issues related to medicine, life sciences and associated technologies as applied to human beings, taking into account their social, legal and environmental dimensions.
2. This Declaration is addressed to States. As appropriate and relevant, it also provides guidance to decisions or practices of individuals, groups, communities, institutions and corporations, public and private.

Article 2—Aims

The aims of this Declaration are:

a. to provide a universal framework of principles and procedures to guide States in the formulation of their legislation, policies or other instruments in the field of bioethics;
b. to guide the actions of individuals, groups, communities, institutions and corporations, public and private;
c. to promote respect for human dignity and protect human rights, by ensuring respect for the life of human beings, and fundamental freedoms, consistent with international human rights law;
d. to recognize the importance of freedom of scientific research and the benefits derived from scientific and technological developments, while stressing the need for such research and developments to occur within the framework of ethical principles set out in this Declaration and to respect human dignity, human rights and fundamental freedoms;
e. to foster multidisciplinary and pluralistic dialogue about bioethical issues between all stakeholders and within society as a whole;
f. to promote equitable access to medical, scientific and technological developments as well as the greatest possible flow and the rapid sharing of knowledge concerning those developments and the sharing of benefits, with particular attention to the needs of developing countries;
g. to safeguard and promote the interests of the present and future generations;
h. to underline the importance of biodiversity and its conservation as a common concern of humankind.

20.1.3 *Principles*

Within the scope of this Declaration, in decisions or practices taken or carried out by those to whom it is addressed, the following principles are to be respected.

Article 3—Human Dignity and Human Rights

1. Human dignity, human rights and fundamental freedoms are to be fully respected.
2. The interests and welfare of the individual should have priority over the sole interest of science or society.

Article 4—Benefit and Harm

In applying and advancing scientific knowledge, medical practice and associated technologies, direct and indirect benefits to patients, research participants and other affected individuals should be maximized and any possible harm to such individuals should be minimized.

Article 5—Autonomy and Individual Responsibility

The autonomy of persons to make decisions, while taking responsibility for those decisions and respecting the autonomy of others, is to be respected. For persons who are not capable of exercising autonomy, special measures are to be taken to protect their rights and interests.

Article 6—Consent

1. Any preventive, diagnostic and therapeutic medical intervention is only to be carried out with the prior, free and informed consent of the person concerned, based on adequate information. The consent should, where appropriate, be express and may be withdrawn by the person concerned at any time and for any reason without disadvantage or prejudice.
2. Scientific research should only be carried out with the prior, free, express and informed consent of the person concerned. The information should be adequate, provided in a comprehensible form and should include modalities for withdrawal of consent. Consent may be withdrawn by the person concerned at any time and for any reason without any disadvantage or prejudice. Exceptions to this principle should be made only in accordance with ethical and legal standards adopted by States, consistent with the principles and provisions set out in this Declaration, in particular in Article 27, and international human rights law.

3. In appropriate cases of research carried out on a group of persons or a community, additional agreement of the legal representatives of the group or community concerned may be sought. In no case should a collective community agreement or the consent of a community leader or other authority substitute for an individual's informed consent.

Article 7—Persons Without the Capacity to Consent

In accordance with domestic law, special protection is to be given to persons who do not have the capacity to consent:

a. authorization for research and medical practice should be obtained in accordance with the best interest of the person concerned and in accordance with domestic law. However, the person concerned should be involved to the greatest extent possible in the decision-making process of consent, as well as that of withdrawing consent;
b. research should only be carried out for his or her direct health benefit, subject to the authorization and the protective conditions prescribed by law, and if there is no research alternative of comparable effectiveness with research participants able to consent. Research which does not have potential direct health benefit should only be undertaken by way of exception, with the utmost restraint, exposing the person only to a minimal risk and minimal burden and, if the research is expected to contribute to the health benefit of other persons in the same category, subject to the conditions prescribed by law and compatible with the protection of the individual's human rights. Refusal of such persons to take part in research should be respected.

Article 8—Respect for Human Vulnerability and Personal Integrity

In applying and advancing scientific knowledge, medical practice and associated technologies, human vulnerability should be taken into account. Individuals and groups of special vulnerability should be protected and the personal integrity of such individuals respected.

Article 9—Privacy and Confidentiality

The privacy of the persons concerned and the confidentiality of their personal information should be respected. To the greatest extent possible, such information should not be used or disclosed for purposes other than those for which it was collected or consented to, consistent with international law, in particular international human rights law.

Article 10—Equality, Justice and Equity

The fundamental equality of all human beings in dignity and rights is to be respected so that they are treated justly and equitably.

Article 11—Non-Discrimination and Non-Stigmatization

No individual or group should be discriminated against or stigmatized on any grounds, in violation of human dignity, human rights and fundamental freedoms.

Article 12—Respect for Cultural Diversity and Pluralism

The importance of cultural diversity and pluralism should be given due regard. However, such considerations are not to be invoked to infringe upon human dignity, human rights and fundamental freedoms, nor upon the principles set out in this Declaration, nor to limit their scope.

Article 13—Solidarity and Cooperation

Solidarity among human beings and international cooperation towards that end are to be encouraged.

Article 14—Social Responsibility and Health

1. The promotion of health and social development for their people is a central purpose of governments that all sectors of society share.
2. Taking into account that the enjoyment of the highest attainable standard of health is one of the fundamental rights of every human being without distinction of race, religion, political belief, economic or social condition, progress in science and technology should advance:

    a. access to quality health care and essential medicines, especially for the health of women and children, because health is essential to life itself and must be considered to be a social and human good;
    b. access to adequate nutrition and water;
    c. improvement of living conditions and the environment;
    d. elimination of the marginalization and the exclusion of persons on the basis of any grounds;
    e. reduction of poverty and illiteracy.

Article 15—Sharing of Benefits

1. Benefits resulting from any scientific research and its applications should be shared with society as a whole and within the international community, in particular with developing countries. In giving effect to this principle, benefits may take any of the following forms:
   a. special and sustainable assistance to, and acknowledgement of, the persons and groups that have taken part in the research;
   b. access to quality health care;
   c. provision of new diagnostic and therapeutic modalities or products stemming from research;
   d. support for health services;
   e. access to scientific and technological knowledge;
   f. capacity-building facilities for research purposes;
   g. other forms of benefit consistent with the principles set out in this Declaration.
2. Benefits should not constitute improper inducements to participate in research.

Article 16—Protecting Future Generations

The impact of life sciences on future generations, including on their genetic constitution, should be given due regard.

Article 17—Protection of the Environment, the Biosphere and Biodiversity

Due regard is to be given to the interconnection between human beings and other forms of life, to the importance of appropriate access and utilization of biological and genetic resources, to respect for traditional knowledge and to the role of human beings in the protection of the environment, the biosphere and biodiversity.

20.1.4 Application of the Principles

Article 18—Decision-Making and Addressing Bioethical Issues

1. Professionalism, honesty, integrity and transparency in decision-making should be promoted, in particular declarations of all conflicts of interest and appropriate sharing of knowledge. Every endeavour should be made to use the best available scientific knowledge and methodology in addressing and periodically reviewing bioethical issues.

2. Persons and professionals concerned and society as a whole should be engaged in dialogue on a regular basis.
3. Opportunities for informed pluralistic public debate, seeking the expression of all relevant opinions, should be promoted.

Article 19—Ethics Committees

Independent, multidisciplinary and pluralist ethics committees should be established, promoted and supported at the appropriate level in order to:

a. assess the relevant ethical, legal, scientific and social issues related to research projects involving human beings;
b. provide advice on ethical problems in clinical settings;
c. assess scientific and technological developments, formulate recommendations and contribute to the preparation of guidelines on issues within the scope of this Declaration;
d. foster debate, education and public awareness of, and engagement in, bioethics.

Article 20—Risk Assessment and Management

Appropriate assessment and adequate management of risk related to medicine, life sciences and associated technologies should be promoted.

Article 21—Transnational Practices

1. States, public and private institutions, and professionals associated with transnational activities should endeavour to ensure that any activity within the scope of this Declaration, undertaken, funded or otherwise pursued in whole or in part in different States, is consistent with the principles set out in this Declaration.
2. When research is undertaken or otherwise pursued in one or more States (the host State(s)) and funded by a source in another State, such research should be the object of an appropriate level of ethical review in the host State(s) and the State in which the funder is located. This review should be based on ethical and legal standards that are consistent with the principles set out in this Declaration.
3. Transnational health research should be responsive to the needs of host countries, and the importance of research contributing to the alleviation of urgent global health problems should be recognized.
4. When negotiating a research agreement, terms for collaboration and agreement on the benefits of research should be established with equal participation by those party to the negotiation.

5. States should take appropriate measures, both at the national and international levels, to combat bioterrorism and illicit traffic in organs, tissues, samples, genetic resources and genetic-related materials.

20.1.5 Promotion of the Declaration

Article 22—Role of States

1. States should take all appropriate measures, whether of a legislative, administrative or other character, to give effect to the principles set out in this Declaration in accordance with international human rights law. Such measures should be supported by action in the spheres of education, training and public information.
2. States should encourage the establishment of independent, multidisciplinary and pluralist ethics committees, as set out in Article 19.

Article 23—Bioethics Education, Training and Information

1. In order to promote the principles set out in this Declaration and to achieve a better understanding of the ethical implications of scientific and technological developments, in particular for young people, States should endeavour to foster bioethics education and training at all levels as well as to encourage information and knowledge dissemination programmes about bioethics.
2. States should encourage the participation of international and regional intergovernmental organizations and international, regional and national non governmental organizations in this endeavour.

Article 24—International Cooperation

1. States should foster international dissemination of scientific information and encourage the free flow and sharing of scientific and technological knowledge.
2. Within the framework of international cooperation, States should promote cultural and scientific cooperation and enter into bilateral and multilateral agreements enabling developing countries to build up their capacity to participate in generating and sharing scientific knowledge, the related know-how and the benefits thereof.
3. States should respect and promote solidarity between and among States, as well as individuals, families, groups and communities, with special regard for those

rendered vulnerable by disease or disability or other personal, societal or environmental conditions and those with the most limited resources.

Article 25—Follow-Up Action by UNESCO

1. UNESCO shall promote and disseminate the principles set out in this Declaration. In doing so, UNESCO should seek the help and assistance of the Intergovernmental Bioethics Committee (IGBC) and the International Bioethics Committee (IBC).
2. UNESCO shall reaffirm its commitment to dealing with bioethics and to promoting collaboration between IGBC and IBC.

20.1.6 Final Provisions

Article 26—Interrelation and Complementarity of the Principles

This Declaration is to be understood as a whole and the principles are to be understood as complementary and interrelated. Each principle is to be considered in the context of the other principles, as appropriate and relevant in the circumstances.

Article 27—Limitations on the Application of the Principles

If the application of the principles of this Declaration is to be limited, it should be by law, including laws in the interests of public safety, for the investigation, detection and prosecution of criminal offences, for the protection of public health or for the protection of the rights and freedoms of others. Any such law needs to be consistent with international human rights law.

Article 28—Denial of Acts Contrary to Human Rights, Fundamental Freedoms and Human Dignity

Nothing in this Declaration may be interpreted as implying for any State, group or person any claim to engage in any activity or to perform any act contrary to human rights, fundamental freedoms and human dignity.

20.2 Appendix 2

20.2.1 *Report of IBC on the Principle of Respect for Human Vulnerability and Personal Integrity*[2]

This Report is the result of a long reflection within the International Bioethics Committee (IBC) which began at its fifteenth session (Paris, October 2008), was pursued at its sixteenth session (Mexico, October 2009), and was further developed within the framework of its work programme for 2010–2011. It does not pretend to be exhaustive nor prescriptive and does not necessarily represent the views of the Member States of UNESCO.

20.2.2 *Introduction*

1. Article 1 of the UNESCO Universal Declaration on Bioethics and Human Rights (2005) (hereafter "the Declaration") states that it seeks to address the "ethical issues related to medicine, life sciences and associated technologies as applied to human beings, taking into account their social, legal and environmental dimensions". Article 8 reinforces this commitment by linking it to respect for personal integrity and the need to protect vulnerable individuals and groups:

   > In applying and advancing scientific knowledge, medical practice and associated technologies, human vulnerability should be taken into account. Individuals and groups of special vulnerability should be protected and the personal integrity of such individuals respected.

2. This notion is not new. The concept of vulnerability appears in important national documents, starting with the US Belmont Report of 1978, and in international documents, such as the third and most complete version of the International Ethical Guidelines for Biomedical Research Involving Human Subjects of the Council for International Organizations of Medical Sciences (CIOMS) (2002) and in the latest (2008) version of the Declaration of Helsinki, which makes specific reference to vulnerability in Articles 9 and 17.
3. Article 8 of the Declaration entails both a 'negative' duty to refrain from doing something and a 'positive' duty to promote solidarity and to share the benefits of scientific progress. There is an integral relationship between respect for the integrity and dignity of persons on the one hand and the vulnerability of persons on the other. Indeed, UNESCO itself had previously acknowledged the importance of the principle according to which States "should respect and

[2] This is an excerpt of nos. 1–21, 40–44, as the cases and examples are left out. The complete text can be found on http://unesdoc.unesco.org/images/0018/001895/189591e.pdf.

promote the practice of solidarity towards individuals, families and population groups who are particularly vulnerable" with regard to disease or disability of a genetic character (Universal Declaration on the Human Genome and Human Rights 1997).

4. The UNESCO International Bioethics Committee (IBC) has focused its two more recent Reports on the principles of consent (Article 6 of the Declaration) and of social responsibility and health (Article 14). This Report on Article 8 of the Declaration will investigate the scope and content of the principle of respect for human vulnerability and personal integrity, focusing on special vulnerabilities and taking into account conditions that, more or less directly, impinge upon the capacity to live as a free, autonomous individual and the right to live in a world where significant inequalities in the capacity to meet everyone's basic needs are adequately addressed.
5. Article 1.2 of the Declaration makes it clear that it is addressed to States. However, States and governments are not the only addressees of Article 8. As in the case of the principle of social responsibility, it is rather necessary to boost awareness of the responsibility that all sectors of society share and to promote, at the international as well as the domestic level, those strategies and means of cooperation that are most likely to effectively address the determinants of "special" vulnerability to which Article 8 refers. Of course, both an in-depth reflection on the concept of vulnerability as a general feature of the human condition and denunciation of political, economic or cultural discrimination among human beings are important. Nonetheless, they constitute the background more than the focal point of the challenge faced in Article 8. The specific task of this Article is to address special vulnerabilities that occur, whether as a consequence of personal disability, environmental burdens or social injustice, in the contexts of health care, research and the application of emerging technologies in the biomedical sciences. Article 8 enjoins everyone to exercise vigilance in protecting the well-being of individuals and groups in these contexts. As the Declaration (taken as a whole) confirms, every human being has a claim to our care that must be respected.

20.2.3 The Determinants of "Special Vulnerability"

1. The human condition implies vulnerability. Every human being is exposed to the permanent risk of suffering "wounds" to their physical and mental integrity. Vulnerability is an inescapable dimension of the life of individuals and the shaping of human relationships. To take into account human vulnerability acknowledges that we all may lack at some point the ability or the means to protect ourselves, our health and our well-being. We are all confronted with the possibility of disease, disability and environmental risks. At the same time, we live with the possibility that harm, even death, can be caused by other human beings.

2. Of course, Article 8 does not require us to protect vulnerability as such, but rather vulnerable individuals, families and groups in the contexts in which they live. While some groups of people can always be considered vulnerable because of their status (e.g. children), others may be vulnerable in one situation but not in another. Therefore, vulnerability cannot be considered as a one-off concept. The principle of respect underpinned in Article 8 entails a commitment to identify threats to well-being and appropriate means to foster the principles stated in Article 3 to be the primary ones "to be respected": human dignity, human rights and fundamental freedoms. Thus, attempts to define vulnerability in general risk drawing the concept too widely or too narrowly, thereby triggering disputes rather than resolving them. In most cases, however, it is relatively easy to recognise vulnerability when it arises: something fundamental is indeed at stake.
3. That is also why human vulnerability and personal integrity, the other essential concept evoked in Article 8, relate to each other. When a part of our body is inappropriately 'touched' (this is the meaning of the ancient Latin verb from which the noun 'integrity' stems), our life itself, or at least our health, may be threatened. When our freedom is hampered, either by adverse circumstances or by the actions of others, we experience a "wound" to our identity, to its value and dignity. Preservation of integrity implies protection against these kinds of intrusions, the capacity to "say no" to any sort of impingement upon our freedom or to any sort of exploitation of our body and our environment. We are nonetheless committed at least to seek to ameliorate the effects of harms and disadvantages imposed by circumstances. This is a prerequisite of human flourishing and self-fulfilment.
4. The Declaration, taken as a whole, enjoins governments, but also public and private institutions and individuals, to exercise greater vigilance in protecting the well-being of every human being in the face of advances in the fields of medicine, life sciences and associated technologies. By doing so, it underscores the importance of a wide array of principles familiar to the language of both moral and juridical sciences: autonomy, beneficence, justice, dignity, equality and the like. The principle of respect for vulnerability and personal integrity not only emphasizes these aims, but also clarifies that the final goal of the progress of science in the bioethical domain cannot solely be profit. Vulnerability as a human condition calls on every human being, especially those who have the responsibility to advance knowledge and to decide how to use it, to fulfil the fundamental obligations we have one to another. It has been said that acknowledging the reality of vulnerability might provide a bridge between the moral 'strangers' of a pluralistic society, thereby enhancing the value of solidarity rather than mere individual interests.
5. At the same time, Article 8 clarifies that we cannot be satisfied with the simple exercise of restraint and forbearance in pursuing our own objectives when this might threaten the autonomy and dignity of others. We are compelled to act in a positive way to help other people cope with the natural or social determinants of vulnerability. Article 24 underscores the duty to "respect and promote solidarity between and among States" and points out some circumstances that may render

individuals, families and groups vulnerable: disease; disability; other personal, societal or environmental conditions or limited resources. There is no doubt that the empowerment of people to protect them from special vulnerabilities in health care entails making more resources available, providing safe living conditions and access to quality health care as a precondition to every human being's "enjoyment of the highest attainable standard of health" (Article 14 of the Declaration). In this sense, commitment to respect for human vulnerability and personal integrity is a necessary constituent of unavoidable political responsibilities.

6. Humankind as such is vulnerable, but there are individuals, groups and situations to which a greater attention must be paid. This is the essential point to underline. Article 8 explicitly addresses the "special vulnerability" of individuals and groups, inasmuch as they are potential recipients of therapies, involved in scientific research or potential recipients of the products or technologies deriving from the advancement and applications of scientific knowledge. Of course, this is not the only context in which vulnerability occurs. Human dignity, human rights and fundamental freedoms are equally at stake in many other fields of human activity. They are, however, relevant to the scope of the Declaration only insofar as they overlap with the specific tasks of bioethics and medical ethics.
7. Two fundamental categories are highlighted that are relevant to these special responsibilities and obligations:

    a. special (temporary or permanent) disabilities, disease and limitations imposed by the stages of human life;
    b. social, political and environmental determinants: for example culture, economy, relations of power, natural disasters.

8. In a) for example, children are assumed to be vulnerable regardless of their social conditions. Elderly people may be more vulnerable because of their reduced physical and sometimes mental capacities. Persons with disabilities need help to access and sustain the exercise of their self-determination. Persons with mental disorders may not be able to defend themselves or claim their rights. These can all be considered as "natural" determinants of special individual vulnerabilities. Of course, a crucial distinction is to be made between these and special vulnerabilities which result from a deliberate restriction of autonomy.
9. The issue of social, political and environmental determinants is more complex and involves the fundamental matter of justice in the relations between individuals, groups and States. Many individuals, groups and populations nowadays become especially vulnerable because of factors created and implemented by other human beings, in many cases in blatant violation of fundamental human rights. Social vulnerability is a phenomenon determined by the structure of people's and communities' daily lives. Situations of social vulnerability usually interfere with the self-determination of individuals and lead to significantly increased exposure to risks caused by social exclusion. Social vulnerability plays a role not only in biomedical research but also in the healthcare setting and in the development, implementation and application of emerging technologies in biomedical sciences and is a fact of life for a considerable portion of world's

population. Vulnerability is caused or exacerbated by a lack of means and of the capacity to protect oneself, as in the following examples:

- poverty, inequalities in income, social conditions, education and access to information (e.g. unemployed persons, homeless persons, illiterate persons, individuals involved in research activities that follow a 'double standard' procedure in which the same research is not subjected to the same ethical scrutiny in different locations);
- gender discrimination (e.g. lack of equitable access to healthcare);
- situations of substantial limitation or deprivation of personal liberty (e.g. prisoners);
- hierarchical relations (e.g. students involved in research projects carried out by their teachers, employees in situations where safe working conditions are not guaranteed, members of the armed forces or the police);
- marginalization on various grounds (e.g. immigrants, nomads, ethnic and racial minorities);
- trade-offs between the right of every human being to quality health care and other rights, such as intellectual property rights, whose inappropriate protection can directly or indirectly impinge negatively upon the sharing of the benefits of scientific progress;
- exploitation of resources in developing countries (e.g. the consequences of deforestation which can compromise duties towards future generations);
- wars (e.g. asylum seekers and displaced persons);
- negative effects of human activity, for example climate change or different kinds of pollution;
- impact of natural disasters like earthquakes, hurricanes or tsunamis leading to death, injury and displacement.

10. All of these examples help contextualize the issue of vulnerability and respect for personal integrity within the three specific domains pointed out in Article 8. They serve to highlight the issues in the healthcare setting, in the researcher-participant relationship in human participant research, and in the development and application of emerging technologies in the biomedical sciences. In each of these settings people are vulnerable in ways over and above that which the human condition necessarily involves.
11. As to vulnerability in the healthcare setting, it is well understood that even those patients whose physical and cognitive capacities meet or exceed those of the average human population are uniquely vulnerable when submitting to medical care, given the greater expertise and social authority of the treating physician (and other professionals). A patient's vulnerability may be further aggravated by his or her illness—pain, discomfort and the desire for healing can frustrate reasoning and sound judgment. This is a fortiori true of those patients whose physical or cognitive abilities are severely diminished such that their capacity for self-determination is limited or even non-existent. In all healthcare settings, the patient, to a greater or lesser degree, depends on the skills, expertise, judgment and good will of the treating professional. Individually and collec-

tively, patients are thereby uniquely vulnerable. Article 8 calls our attention to this fact and enjoins decision makers to respond in a fitting way. One possible option would be for States to develop a patients' rights charter which would be instrumental in protecting those who are especially vulnerable in the healthcare setting.

12. The same is true of human participants in biomedical research. However, in this context, the additional safeguard of the physician's (or researcher's) primary focus being on restoring the patient to health is absent. The interests of researchers and subjects are not always aligned as they are in the relationship between healthcare professionals and patients, thus amplifying concerns regarding vulnerability in this domain. Moreover, in some instances, the researcher's (or society's) understandably strong desire to pursue useful generalisable knowledge gives rise to the temptation to under-protect or ignore the participants' well being. The Declaration's emphasis on special vulnerability rightly invites careful attention to this and related concerns in the human research context. It is clear, however, that the engagement of people as participants in clinical research is key in providing solutions to, and understanding of, medical problems afflicting humankind. Biomedical research, including clinical research, has evolved over the years to the extent that international and national standards and guidelines have been developed. The practices uncovered at the Nuremberg trials showed the range and extent of the abuse of human beings in research and resulted in a flurry of activity on this subject. The resulting Nuremberg Code [3] provided the template for a number of successive declarations on human participant research, culminating in the World Medical Association's Declaration of Helsinki, which states that:

> Medical research is subject to ethical standards that promote respect for all human subjects and protect their health and rights. Some research populations are particularly vulnerable and need special protection. These include those who cannot give or refuse consent themselves and those who may be vulnerable to coercion or undue influence.[4]

13. Equally, some individuals or communities may be denied access to, or the benefits of, research.They too are especially vulnerable.
14. It is especially important in all forms of research that close attention is paid to the kinds of pressures that might encourage people to act in a manner that potentially competes with their own interests; in particular, the avoidance of coercion or of the appearance of coercion is vital. Coercion may arise from a number of sources, including the simple omission of relevant information about possible risk. As research is often conducted by investigators from the developed world in countries of the developing world, a number of concerns arise acutely. First, the personal, economic or socio-political situation of potential research participants may render them vulnerable to exploitation. Second,

[3] Available at http://ohsr.od.nih.gov/guidelines/nuremberg.html. Accessed 03/08/2010.

[4] Declaration of Helsinki, Article 9. Most recently revised at the 59th WMA General Assembly, Seoul, October 2008.

again because of the so-called 'therapeutic misconception', people may agree to participate in research in the mistaken belief that there may be some benefit for them; this is particularly likely where healthcare services are inadequate or unavailable. CIOMS' most recent guidelines note that, special justification is required for inviting vulnerable individuals to serve as research participants and, if they are selected, the means of protecting their rights and welfare must be strictly applied[5].

15. Advances in biomedical science and biotechnology have opened the way for many new and powerful capacities for the protection of human welfare. At the same time, they have created novel and potent mechanisms of exploitation and degradation. Such advances potentially give rise to a special vulnerability for individuals and groups to which Article 8 rightly draws our attention.
16. Each of these contexts presents instances where individuals and groups may be vulnerable in a unique and amplified form. A number of examples follow which illuminate some situations within healthcare, human participant research and the development of biotechnology which give rise to "special vulnerability". It is, of course, not intended that these examples are exhaustive of situations where vulnerability can be identified; rather, they are illustrative.

20.2.4 Conclusions

1. In this report, IBC has provided both a theoretical account of the principle of respect for personal integrity and the need to protect those who are especially vulnerable, as well as a series of practical examples. These examples are not exhaustive of the issues that could be raised; they are rather intended as a useful template for further discussion and development.
2. Vulnerability as a risk of a human being to be harmed in his or her physical and mental integrity is an element of human condition. Special vulnerability in the scope of Article 8 of the Universal Declaration on Bioethics and Human Rights means that there are individuals and groups that are especially prone to violation of personal integrity or disrespect for autonomy due to exploitation, deception, coercion and disregard through the application and advancing of scientific knowledge, medical practice and associated technologies. People can be especially vulnerable for many reasons: because of their age like children, their kind of disease like rare or neglected diseases, because of lacking access to health care due to the health care system of their country, their own education or the education of physicians and researchers.
3. A further important example of special vulnerabilities can be provided using the example of the position of women. In some cultures, female children are uniquely vulnerable to the risk of being unwanted, uncared for, abused and

[5] Macrae (2007).

rejected. Female children may also find their interest in bodily integrity gravely threatened, including especially their right to be free from sexual assault and exploitation. Adult women may find themselves transferred from the patronage of their father to the patronage of their husband, thereby denying them the personal authority to make important life and healthcare decisions on their own behalf. As women live longer in many parts of the world, elderly women might find themselves abandoned by their families, subject to inadequate healthcare, and disregarded by society. Migrant women and women affected by war are especially vulnerable to abuse and are often disenfranchised from engaging in conflict resolution and reconciliation.

4. It must be accepted that situations of vulnerability seldom exist in isolation. Lack of access to education, lack of social authority, limited access to healthcare and freedom from coercion can combine negatively to affect the integrity of people throughout the world. In addition, there are often complex, social, cultural and political barriers that negatively impact on respect for personal integrity, and create seemingly intractable, situations of special vulnerability for both individuals and groups. In particular, lower levels of education always predict higher levels of vulnerability.
5. In its Article 1, the Universal Declaration on Bioethics and Human Rights makes it clear that it is addressed to States, but also to individuals, groups, communities, institutions and corporations, public or private. We all share responsibilities in this area. While we cannot eradicate vulnerability entirely, given that it is a feature of the human condition, we can and should provide every human being with the best available means to ensure that they do not find themselves in a position of special vulnerability, regardless of age, gender, educational level, financial situation, health status and life experiences. Securing the protection of groups and individuals with special vulnerability, by addressing the context and causes that give rise to it, is the foremost test of our capacity and willingness to foster the idea of equal rights and the dignity of every human being.

References

Macrae, D. J. 2007. The Council for International Organizations and Medical Sciences (CIOMS) guidelines on ethics of clinical trials. *Proceedings of the American Thoracic Society* 4:176–179.

Index

J. Tham et al. (eds.), *Religious Perspectives on Human Vulnerability in Bioethics*, Advancing Global Bioethics 2, DOI 10.1007/978-94-017-8736-9,

Printed by Printforce, the Netherlands